NIST Special Publication 250-41

NIST MEASUREMENT SERVICES:
Spectroradiometric Detector Measurements:
Part I–Ultraviolet Detectors and Part II–Visible to Near-Infrared Detectors

Thomas C. Larason, Sally S. Bruce, and Albert C. Parr

Optical Technology Division
Physics Laboratory
National Institute of Standards and Technology
Gaithersburg, MD 20899-0001

Supersedes NBS Special Publication 250-17

February 1998

U.S. DEPARTMENT OF COMMERCE
William M. Daley, Secretary

Technology Administration
Gary R. Bachula, Acting Under Secretary for Technology

National Institute of Standards and Technology
Raymond G. Kammer, Director

Preface

The calibration and related measurement services of the National Institute of Standards and Technology (NIST) are intended to assist the makers and users of precision measuring instruments in achieving the highest possible levels of accuracy, quality, and productivity. NIST offers over 300 different calibrations, special tests, and measurement assurance services. These services allow customers to directly link their measurement systems to measurement systems and standards maintained by NIST. NIST offers these services to the public and private organizations alike. They are described in NIST Special Publication (SP) 250, *NIST Calibration Services Users Guide*.

The Users Guide is supplemented by a number of Special Publications (designated as the "SP250 Series") that provide detailed descriptions of the important features of specific NIST calibration services. These documents provide a description of the: (1) specifications for the services; (2) design philosophy and theory; (3) NIST measurement system; (4) NIST operational procedures; (5) assessment of the measurement uncertainty including random and systematic errors and an error budget; and (6) internal quality control procedures used by NIST. These documents will present more detail than can be given in NIST calibration reports, or than is generally allowed in articles in scientific journals. In the past, NIST has published such information in a variety of ways. This series will make this type of information more readily available to the user.

This document, SP250-41 (1998), NIST Measurement Services: Spectroradiometric Detector Measurements, is a revision of SP250-17 (1988). It covers the calibration of standards and special tests of photodetector absolute spectral responsivity from 200 nm to 1800 nm (Service ID numbers 39071S - 39081S in SP250, NIST Calibration Services Users Guide). Inquiries concerning the technical content of this document or the specifications for these services should be directed to the authors or to one of the technical contacts cited.

NIST welcomes suggestions on how publications such as this might be made more useful. Suggestions are also welcome concerning the need for new calibration services, special tests, and measurement assurance programs.

Peter L. M. Heydemann (Acting)
Director
Measurement Services

Katharine B. Gebbie
Director
Physics Laboratory

Abstract and Key Words

The National Institute of Standards and Technology supplies calibrated photodiode standards and special tests of photodetectors for absolute spectral responsivity from 200 nm to 1800 nm. (This service will soon be expanded to 20 µm in the infrared.) The scale of absolute spectral responsivity is based solely on detector measurements traceable to the High Accuracy Cryogenic Radiometer maintained by the National Institute of Standards and Technology. Silicon photodiode light-trapping detectors are used to transfer the optical power unit from this cryogenic radiometer to monochromator-based facilities where routine measurements are performed. The transfer also involves modeling the quantum efficiency of the silicon photodiode light-trapping detectors. A description of current measurement services is given along with the procedures, equipment, and techniques used to perform these calibrations. Detailed estimates and procedures for determining uncertainties of the reported values are also presented.

Key Words: absolute spectral responsivity; calibration; cryogenic radiometer; light-trapping detectors; optical measurement; optical power; photodetector; photodiode; quantum efficiency; quality system; radiometry; responsivity; scale; standards; silicon photodiode

TABLE OF CONTENTS

1. INTRODUCTION 1

2. NIST SPECTRORADIOMETRIC DETECTOR MEASUREMENT SERVICE 3
2.1 Description of Measurement Services 3
2.2 Measurement Limitations 6
2.3 How To Order Photodiodes Or Special Tests 6
2.4 Technical Contacts 7

3. MEASUREMENT THEORY 8
3.1 Measurement Equation 8
    3.1.1 Approximations 11
3.2 Substitution Method 13
    3.2.1 General Substitution Method 13
    3.2.2 Photodetector Substitution 13
    3.2.3 Substitution Method with Monitor 14
    3.2.4 Measurement Equation Applied to the SCFs 14

4. EQUIPMENT DESCRIPTION 16
4.1 Visible to Near-Infrared (Vis/NIR) Comparator Description 16
    4.1.1 Vis/NIR Source 17
    4.1.2 Vis/NIR Monochromator 18
    4.1.3 Vis/NIR Optics 18
    4.1.4 Vis/NIR Translation Stages - Detector Positioning 19
    4.1.5 Vis/NIR Working Standards 19
    4.1.6 Beam Splitter and Monitor Detector 19
    4.1.7 Alignment Lasers 19
    4.1.8 Enclosure 19
4.2 Ultraviolet (UV) Comparator Description 20
    4.2.1 UV Source 20
    4.2.2 UV Monochromator 21
    4.2.3 UV Optics 21
    4.2.4 UV Translation Stages - Detector Positioning 21
    4.2.5 UV Working Standards 22
    4.2.6 Beam Splitter and Monitor Detector 22
    4.2.7 Alignment Lasers 22
    4.2.8 Enclosure 22
4.3 Electronics 22
    4.3.1 Electronics - Signal Measurement 22
    4.3.2 Electronics - Auxiliary Equipment 25

5. ABSOLUTE SPECTRAL RESPONSIVITY SCALE REALIZATION 25
5.1 Transfer from HACR to Traps (405 nm to 920 nm) 25
5.2 Traps to Visible Silicon Working Standards 28
5.3 Extension to Ultraviolet (200 nm) and Near-Infrared (1800 nm) 29

    5.3.1 Pyroelectric Detector 29
    5.3.2 Responsivity Measurements 30
5.4 Germanium (NIR) Working Standards 30
5.5 UV Silicon Working Standards 30
5.6 Extension of Visible Silicon Working Standards 31

6. CALIBRATION PROCEDURES AND COMPUTER AUTOMATION 33

6.1 Spectral Responsivity 33
    6.1.1 Calibration Procedures 33
    6.1.2 Quantum Efficiency 35
6.2 Spatial Uniformity 36
    6.2.1 Measurement Method and Calibration Procedure 37
    6.2.2 Limitations 37
6.3 Computer Automation 40
    6.3.1 Computer Automated Equipment 40
    6.3.2 Computer Calibration Programs 43

7. UNCERTAINTY ASSESSMENT 46

7.1 Uncertainty Components 46
    7.1.1 "Indirect" Uncertainty Components 49
    7.1.2 Uncertainty Components due to Assumptions and Approximations 51
    7.1.3 Other Factors Considered and Neglected 54
7.2 Transfer from Traps (HACR) to Working Standards 55
    7.2.1 Visible Silicon Working Standards 55
    7.2.2 Germanium (NIR) Working Standards 57
    7.2.3 UV Silicon Working Standards 58
7.3 Transfer to Test (Customer) Detectors 60
    7.3.1 UV Silicon Transfer 61
    7.3.2 Visible Silicon Transfer 61
    7.3.3 NIR Transfer 62
    7.3.4 Filtered Detector Transfer 64
7.4 Spatial Uniformity Measurement Uncertainty 64

8. QUALITY SYSTEM 65

8.1 Control Charts 65
8.2 Comparison to Other Laboratories 66

9. CHARACTERISTICS OF PHOTODIODES AVAILABLE FROM NIST 67

9.1 Hamamatsu S1337-1010BQ 67
9.2 Hamamatsu S2281 68
9.3 UDT Sensors UV100 68
9.4 Detector Apertures 70
9.5 Detector Fixture Mechanical Drawings 70

10. FUTURE WORK 74

11. ACKNOWLEDGMENTS 74

12. REFERENCES 75

13. BIBLIOGRAPHY 82

APPENDIX - SAMPLE TEST REPORTS

39071S - UV Silicon Photodiode Spectral Responsivity Test Report Sample A-1

39073S - Visible to Near IR Silicon Photodiode Spectral Responsivity Test Report Sample A-8

39075S - Near IR Photodiode Spectral Responsivity Test Report Sample A-15

39081S - Responsivity Spatial Uniformity Test Report Sample A-21

LIST OF FIGURES

Figure 2.1. NIST UV, visible, and near-IR spectral responsivity measurement uncertainties 4
Figure 3.1. The geometry for detector spectral responsivity measurements 9
Figure 3.2. Block diagram of photodetector substitution method 14
Figure 3.3. Block diagram of photodetector substitution method with monitor 14
Figure 4.1. Visible to Near-Infrared Spectral Comparator Facility (Vis/NIR SCF) 17
Figure 4.2. Spectral output flux of the UV and visible to near-IR monochromators 18
Figure 4.3. Ultraviolet Spectral Comparator Facility (UV SCF) 20
Figure 4.4. Vis/NIR detector, amplifier, DVM, and computer control block diagram 23
Figure 4.5. NIST SCF precision transimpedance (I/V) amplifier circuit 24
Figure 4.6. Vis/NIR detector, amplifier, lock-in, DVM, and computer control block diagram 24
Figure 5.1. NIST High Accuracy Cryogenic Radiometer (HACR) 26
Figure 5.2. Trap detector arrangement of photodiodes minimizes light lost to reflections 27
Figure 5.3. Scale transfer by substitution method with the HACR 27
Figure 5.4. NIST spectral power scale propagation chain 28
Figure 5.5. NIST detector spectral responsivity scale realization 32
Figure 6.1. Typical signals from Hamamatsu S1337 and monitor photodiodes 34
Figure 6.2. Spectral responsivities of typical Si, InGaAs, and Ge photodiodes 35
Figure 6.3. Quantum efficiencies of typical Si, InGaAs, and Ge photodiodes 36
Figure 6.4. Spatial uniformities of typical Hamamatsu S1337 and S2281 photodiodes 38
Figure 6.5. Spatial uniformities of typical UDT Sensors UV100 and EG&G Judson Ge photodiodes 39
Figure 6.6. Vis/NIR SCF computer control block diagram 41
Figure 6.7. UV SCF computer control block diagram 42
Figure 6.8. Measurement (computer) program flowchart for one detector scan 44

Figure 7.1. Visible Working Standard (Vis WS) relative combined standard uncertainty 57
Figure 7.2. Germanium (NIR) Working Standard (Ge WS) relative combined standard uncertainty 57
Figure 7.3. Ultraviolet Working Standard (UV WS) relative combined standard uncertainty 59
Figure 7.4. Transfer to test (customer) detectors relative combined standard uncertainty 61
Figure 8.1. Control chart example for a NIST Visible Working Standard (Vis WS) 66
Figure 9.1. Temperature coefficient of silicon Hamamatsu S1226 and S1337 photodiodes 68
Figure 9.2. Linearity of Hamamatsu S1337-1010BQ at 633 nm 68
Figure 9.3. Linearity of UDT Sensors UV100 at 442 nm 69
Figure 9.4. Responsivity dependence on bias voltage of UDT Sensors UV100 at 442 nm 69
Figure 9.5. Mechanical diagram of Hamamatsu S1337-1010BQ fixture body 71
Figure 9.6. Mounting pieces for the Hamamatsu S1337-1010BQ photodiode and BNC connector 71
Figure 9.7. Exploded view diagram of Hamamatsu S1337-1010BQ fixture 72
Figure 9.8. Mechanical diagram of UDT Sensors UV100 and Hamamatsu S2281 fixture 72
Figure 9.9. Mechanical diagram of pre-1993 aperture plate for detector fixtures 73
Figure 9.10. Mechanical diagram of present aperture plate for detector fixtures 73

LIST OF TABLES

Table 2.1. NIST Spectroradiometric Detector Measurement Services 3
Table 2.2. Detector Measurement Services Uncertainties 4
Table 7.1. Visible Working Standard Uncertainty 56
Table 7.2. Germanium Working Standard Uncertainty 58
Table 7.3. UV Working Standard Uncertainty 59
Table 7.4. Transfer Uncertainty to Test (Customer) UV100 Silicon Photodiodes 61
Table 7.5. Transfer Uncertainty to Test (Customer) S1337 and S2281 Silicon Photodiodes 62
Table 7.6. Transfer Uncertainty to Test (Customer) TE Cooled Germanium Photodiodes 63
Table 7.7. Transfer Uncertainty to Test (Customer) InGaAs Photodiodes 64
Table 7.8. Photodetector Spatial Uniformity Measurement Repeatablity Uncertainty 65

1. Introduction

This document describes the National Institute of Standards and Technology (NIST) Measurement Service for providing calibrated photodiodes for absolute spectral responsivity in the ultraviolet (UV), visible, and near-infrared (NIR) spectral regions (200 nm to 1100 nm) and the special service of measuring photodetectors sent to NIST in the 200 nm to 1800 nm spectral region. This document supersedes NBS Special Publication 250-17 (1988), "The NBS Photodetector Spectral Response Calibration Transfer Program."

The theory, measurement system, operation, and transfer standards of the Spectroradiometric Detector Measurement Service are described in this publication. The traceability of the absolute spectral power responsivity scale to the High Accuracy Cryogenic Radiometer (HACR) and detailed uncertainty estimates are discussed. Also presented are the recently expanded list of absolute spectral power responsivity measurement services provided by NIST and the enhancements to its quality system to comply with ANSI/NCSL Z540-1-1994 [1].

The material presented in this document describes the equipment and procedures for the Spectroradiometric Detector Measurement Service as they exist at the time of publication. Improvements in the equipment, procedures, and services offered are continuous. The discussions in this document will be primarily directed at the procedures developed to measure the absolute responsivity of photodiodes supplied by NIST to customers. The procedures for characterizing customer-supplied detectors is a straightforward extension of the underlying procedures developed for the NIST furnished devices.

Note: This document follows the NIST policy of using the International System of Units (SI). Only units of the SI and those units recognized for use with the SI are used. Equivalent values in other units may be given in parentheses following the SI values. The mechanical drawings in section 9.5 were originally prepared in English units and are presented without converting the values shown to SI units.

Background

Many radiometric, photometric, and colorimetric applications require the determination of the absolute spectral power responsivity of photodetectors. The absolute spectral power responsivity is the ratio of the photodetector's signal (amperes or volts) to the spectral radiant flux (watts) incident on the photodetector. The absolute spectral power responsivity is also referred to simply as the absolute spectral responsivity. Accurate measurement of absolute spectral power responsivity of photodetectors has been a service provided by the Optical Technology Division and its predecessors for over 20 years.

Various techniques have been employed to determine photodetector absolute spectral power responsivity [2, 3]. In the late 1970's a room-temperature electrical substitution radiometer (ESR), also known as an electrically calibrated radiometer (ECR), was used in conjunction with lasers [4, 5] as the detector scale base. Relative uncertainties were reported on the order of 1.5 % to 5 % (3 standard deviation estimate[1]) over the spectral range from 390 nm to 1100 nm [6].

> Note: This document follows, to the extent possible, the ISO Guide to the Expression of Uncertainty in Measurement (International Organization for Standardization, Geneva, Switzerland, 1993). Since 1994, the NIST policy has been to conform to the *Guide* in reporting its activities, using an expanded uncertainty coverage factor (as defined in the *Guide*) of $k = 2$. See Ref. [7] for a detailed explanation of the NIST policy. Unless otherwise noted all uncertainties will be stated as $k = 2$.

A major advance came in the early 1980's with the silicon photodiode self-calibration techniques [8, 9] and the subsequent introduction of 100 % quantum efficient (QE) photodiodes [10] and their use as the basis for detector calibrations. These photodiodes were used in the development of the United Detector Technologies (UDT) QED-200[2] trap detector [11], which is a light-trapping device constructed of multiple, windowless, 100 % QE silicon photodiodes. The QED-200 had a limited spectral range (usually 400 nm to 750 nm) where it operated with 100 % QE, and suffered from limited dynamic range due to the relatively high bias currents used. The absolute spectral power responsivity was transferred to customers at this time via the Detector Response Transfer and Intercomparison Program (DRTIP). Customers would rent a radiometer from NIST and transfer the detector scale to their working standard(s). The scale relative uncertainty was typically 0.8 % to 6.0 % (3 standard deviation estimate) over the spectral range from 250 nm to 1100 nm [12].

A second generation trap detector (very similar to the commercially available Graseby Optronics QED-150) was later used as the basis for the NIST detector scale [13]. Both second generation trap detectors are constructed with a different type of silicon photodiode (Hamamatsu S1337-1010). They do not have 100 % QE, but the QE could be measured with the QED-200, and extrapolated with high accuracy over a large spectral range from 400 nm to 900 nm [14]. Relative uncertainties were reported on the order of 0.33 % to 1.0 % (3 standard deviation estimate) over the spectral range from 250 nm to 1100 nm (between 200 nm and 250 nm the relative uncertainty was reported as 5.0 %). The next advance came in the late 1980's when cryogenic ESRs were reported with improved uncertainties over the 100 % QE detector-based measurements [15, 16, 17]. The current state of cryogenic ESRs and future direction of detector-based radiometry are discussed in Refs. [18 and 19].

NIST has recently made changes to improve and expand the absolute spectral responsivity measurements it provides to its customers. The most significant change is to base the

---

[1]The term "3 standard deviation estimate" or "3σ" in the context of this paper is the estimated standard deviation multiplied by 3. Prior work at NIST was generally reported with "3σ" uncertainties.

[2]Certain commercial equipment, instruments, or materials are identified in this paper to foster understanding. Such identification does not imply recommendation or endorsement by the National Institute of Standards and Technology, nor does it imply that the materials or equipment identified are necessarily the best available for the purpose.

measurements on the NIST High Accuracy Cryogenic Radiometer [20]. The HACR is the U.S. primary standard for optical power. The scale of absolute spectral responsivity [21] is transferred by silicon photodiode trap detectors to working standards used with the NIST Visible to Near-Infrared [22] and Ultraviolet Spectral Comparator Facilities (hereafter referred to as the Vis/NIR SCF and UV SCF respectively in this document) where the Spectroradiometric Detector measurements are performed.

2. NIST Spectroradiometric Detector Measurement Service

This section describes the photodetector calibrations and measurements offered by the Optical Technology Division. A complete listing of the calibration services across all the measurement parameters offered by NIST can be found in the NIST Calibration Services Users Guide [23]. Current fees can be found in the Fee Schedule [24] which is updated annually and on the internet at the URL address: http://ts.nist.gov/ts/htdocs/230/233/233.htm.

2.1 Description of Measurement Services

There are two types of measurement services provided by NIST: Fixed Services and Special Tests. Fixed services (Service ID numbers ending in the letter C) have fixed measurement conditions and NIST issues a calibration report to the customer. Special tests (Service ID numbers ending in the letter S) have no fixed measurement conditions; are services that are under development; and/or are for unique customer-supplied test items. Service ID numbers 39071S through 39075S are currently in the process of being converted from Special Tests to Fixed Services (i.e., calibrations).

The present spectral range for photodetector absolute spectral responsivity measurements are from 200 nm to 1800 nm. Table 2.1 lists the services offered with typical measurement ranges and typical uncertainties. All services listed are provided routinely. The relative expanded uncertainties of the Spectroradiometric Detector Measurement Service are shown in figure 2.1 and table 2.2. See section 7 for a detailed explanation of the uncertainties.

Table 2.1. NIST Spectroradiometric Detector Measurement Services

Service ID number	Item of test	Range	Relative expanded uncertainty ($k = 2$)
39071S	Ultraviolet Silicon Photodiodes (UDT UV100)	200 nm to 500 nm	0.4 % to 13 %
39072S	Retest of Ultraviolet Silicon Photodiodes (UDT UV100)	200 nm to 500 nm	0.4 % to 13 %
39073S	Visible to Near-Infrared Silicon Photodiodes (Hamamatsu S2281)	350 nm to 1100 nm Can be extended to 200 nm	0.2 % to 4 % 0.2 % to 13 %
39074S	Retest of Visible to Near-Infrared Silicon Photodiodes (Hamamatsu S1337-1010BQ or S2281)	350 nm to 1100 nm Can be extended to 200 nm	0.2 % to 4 % 0.2 % to 13 %
39075S	Special Tests of Near-Infrared Photodiodes	700 nm to 1800 nm	0.5 % to 7 %[†]
39080S	Special Tests of Radiometric Detectors	200 nm to 1800 nm	0.2 % to 13 %[†]
39081S	Special Tests of Photodetector Responsivity Spatial Uniformity	200 nm to 1800 nm	0.0024 % to 0.05 %[†]

[†]Depends on photodetector and signal level.

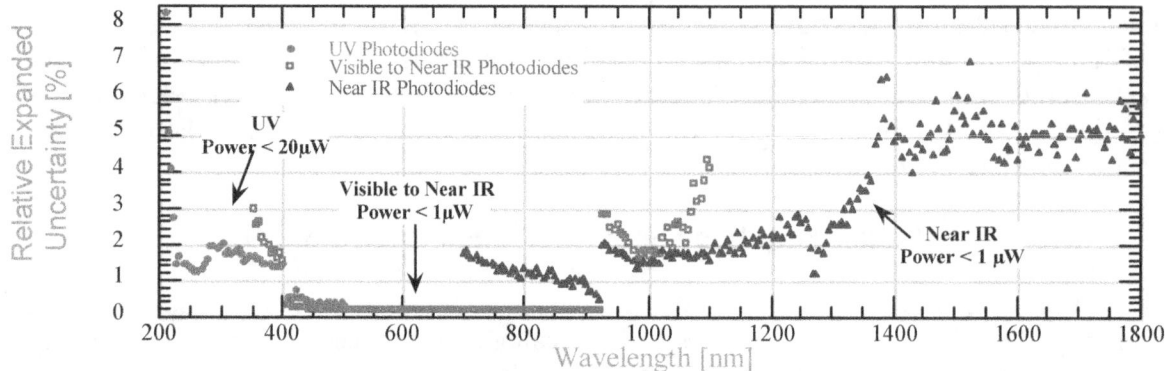

(Note: Relative expanded uncertainty at 200 nm is 13 %.)

Figure 2.1. NIST UV, visible, and near-IR spectral responsivity measurement uncertainties. The three curves are the relative expanded uncertainty ($k = 2$) for measurements with the three different working standard types.

Table 2.2. Detector Measurement Services Uncertainties

Wavelength [nm]	Relative expanded uncertainty ($k = 2$) [%]			
	UV100 (UV)	S1337 (Visible)	Ge (NIR)	InGaAs (NIR)
200	13.02			
250	1.36			
300	2.06			
350	1.68	2.96		
400	1.46	1.56		
450	0.38	0.24		
500	0.38	0.22		
550		0.20		
600		0.20		
650		0.20		
700		0.20	1.78	1.80
750		0.22	1.42	1.46
800		0.22	1.32	1.38
850		0.22	1.12	1.16
900		0.22	0.92	0.96
950		2.58	1.80	1.82
1000		1.72	1.50	1.52
1050		2.66	1.74	1.74
1100		4.16	1.56	1.56
1150			1.92	1.92
1200			2.26	2.26
1250			2.60	2.60
1300			2.56	2.56
1350			3.46	3.46
1400			4.80	4.80
1450			4.64	4.64
1500			5.66	5.66
1550			4.94	4.96
1600			4.30	4.34
1650			5.04	5.08
1700			4.92	5.04
1750			5.24	5.38
1800			5.02	5.36

Descriptions of each Service ID number are provided below.

39071S - UV Silicon Photodiodes
NIST will supply customers with a UDT Sensors, Inc. model UV100 windowed silicon photodiode characterized in the ultraviolet (UV) spectral region. The UV silicon photodiode includes the measured spectral responsivity $[A \cdot W^{-1}]$[3] from 200 nm to 500 nm in 5 nm steps. The 1 cm^2 photosensitive area of the photodiodes is underfilled for the measurements with a beam of diameter 1.5 mm. The spectral responsivity is measured at radiant power levels of less than 20 µW. The bandpass of the measurement is 4 nm. The relative expanded uncertainty ranges from 0.4 % to 13 %, depending on the wavelength. The spatial uniformity of responsivity over the photosensitive area is also measured at 350 nm.

39072S - Retest of UV Silicon Photodiodes
Special tests of UV silicon photodiodes previously supplied by NIST (under 39071S) are performed by measuring spectral responsivity from 200 nm to 500 nm.

39073S - Visible to Near-Infrared Silicon Photodiodes
NIST will supply customers with a Hamamatsu model S2281 (previously a Hamamatsu S1337-1010BQ) windowed silicon photodiode characterized in the visible to near-IR spectral region. The spectral responsivity of the photodiode is measured from 350 nm to 1100 nm in 5 nm steps. The 1 cm^2 photosensitive area of the photodiodes is underfilled for the measurements with a beam of diameter 1.1 mm. The spectral responsivity is measured at radiant power levels of less than 1 µW. The bandpass of the measurement is 4 nm. The relative expanded uncertainty ranges from 0.2 % to 4 %, depending on the wavelength. The spectral range can be extended to 200 nm with a relative expanded uncertainty from 0.2 % to 13 % for an additional fee. The spatial uniformity of responsivity over the photosensitive area is also measured at 500 nm.

39074S - Retest of Visible to Near-Infrared Silicon Photodiodes
Special tests of visible to near-infrared silicon photodiodes previously supplied by NIST (under 39073S) are performed by measuring spectral responsivity from 350 nm to 1100 nm. The spectral range can be extended to 200 nm for an additional fee.

39075S - Special Tests of Near-Infrared Photodiodes
Special tests of customer-supplied near-infrared photodiodes are performed by measuring spectral responsivity from 700 nm to 1800 nm. A beam of diameter 1.1 mm is centered on and underfills the photosensitive area. The spectral responsivity is measured at radiant power levels of less than 1 µW. The bandpass of the measurement is 4 nm. The relative expanded uncertainty ranges from 0.5 % to 7 % or greater, depending on the wavelength and the individual item measured. Customers should communicate with one of the technical contacts listed in section 2.4 to discuss details before submitting a formal request.

---

[3] For clarity, the coherent SI unit is given in brackets for many quantities used in this publication. Of course, SI multiples and submultiples of these units may also be used.

39080S - Special Tests of Radiometric Detectors

Special tests of radiometric detectors in the ultraviolet, visible, and near-infrared regions of the spectrum can be performed. Detector characteristics that can be determined in this special test include spectral responsivity and quantum efficiency (electrons per photon). For example detector responsivity can be measured between 200 nm and 1800 nm at power levels less than 4.0 µW. The relative expanded uncertainty ranges from 0.2 % to 13 % or greater, depending on the wavelength and the individual item measured. Since special tests of this type are unique, details of the tests should be discussed with one of the technical contacts listed in section 2.4 before submitting a formal request.

39081S - Special Tests of Photodetector Responsivity Spatial Uniformity

Special tests consisting of measuring the relative changes in responsivity across the photosensitive area (spatial uniformity) can be performed for customer-supplied photodetectors. The uniformity is typically measured at a single wavelength in 0.5 mm spatial increments with a beam diameter of 1.5 mm in the 200 nm to 400 nm spectral region at power levels less than 20 µW, and a beam of diameter 1.1 mm in the 400 nm to 1800 nm spectral region at power levels less than 1 µW. The relative expanded uncertainty ranges from 0.0024 % to 0.05 % or greater, depending on the wavelength and the individual item measured. Customers should communicate with one of the technical contacts listed in section 2.4 to discuss details before submitting a formal request.

2.2 Measurement Limitations

There are a few limitations on the types of photodetectors that can be measured. Because of the beam size of the comparators, the detector's active area must be greater than 3 mm in diameter. Due to the monochromator flux level an amplifier gain of 10^5 to 10^7 is typically required, thus the photodiode dynamic impedance (shunt resistance) must be greater than 10 kΩ.

Physical size and weight are limited by the translation stages used in the UV SCF and Vis/NIR SCF. Detector packages submitted for testing are limited in size to approximately 20 cm by 20 cm by 20 cm and 2 kg.

Special tests with conditions greatly differing from those listed above may be accepted, but will take longer to complete. The photodetector signal (either voltage or current) must be provided via a BNC connector. Computer communication to a customer's detector package is not possible at this time.

2.3 How To Order Photodiodes Or Special Tests

1) Reference the Service ID number(s) on the purchase order.

   Details of Service ID numbers 39075S, 39080S, and 39081S should be discussed with one of the technical contacts listed in section 2.4 prior to submitting a formal request.

2) The purchase order should be sent to:

National Institute of Standards and Technology
Calibration Program
Building 820, Room 232
Gaithersburg, MD USA 20899-0001

Phone number:	(301) 975-2002
FAX Number:	(301) 869-3548
E-mail:	calibrations@nist.gov
Internet URL:	http://ts.nist.gov/ts/htdocs/230/233/233.htm

3) The purchase order must include the following:

- A) Bill to address.
- B) Ship to address (for the test detector(s) return delivery).
- C) Method of return shipment (the costs below do <u>not</u> include shipping costs). If nothing is stated, NIST will return by common carrier, collect, and uninsured.
- D) User's name, address (for test report), and phone number.
- E) Service ID number(s).

4) The cost for special tests is based on the actual labor and material costs involved and customers are responsible for all shipping costs.

2.4 Technical Contacts

For technical information or questions contact:

Thomas Larason	(301) 975-2334	email: thomas.larason@nist.gov
Sally Bruce	(301) 975-2323	email: sally.bruce@nist.gov
or fax	(301) 869-5700	

Technical information can also be found at the following Internet URL:

http://physics.nist.gov/Divisions/Div844/facilities/phdet/phdet.html

All test detector(s) should be shipped to the following address:

Sally Bruce
NIST
Building 221 / Room B208
Gaithersburg, MD USA 20899-0001

3. Measurement Theory

This section describes the theory and the mathematical basis for the measurement methods used for the services outlined in this publication. The measurement equation is developed and the generalizing assumptions are discussed. The measurement equation is then used in estimating the uncertainties in section 7. The calibration method using detector substitution is described first in general, and then, in detail for the measurement services.

3.1 Measurement Equation

Developing the measurement equation is fundamental to understanding the physics and optics involved with the measurement. This derivation provides an analytical foundation for the measurement process and the assumptions and approximations used in the measurement process. The measurement equation is also fundamental to the analysis of the uncertainty of the measurement process. In essence, a detector comparator is a spectroradiometer that measures the response of different detectors instead of different sources. Thus, the measurement equation developed for a general spectroradiometer can be adapted to detector spectral responsivity measurements.

The measurement equation presented here for spectral responsivity is developed following the general procedure described in chapter 5 of the Self-Study Manual on Optical Radiation Measurements [25] and is similar to the development of eq (7.18) in Ref. [26] for spectral irradiance when using a monochromator-based spectroradiometer. The measurement equation for spectral responsivity is

$$V(A, \Delta\lambda, \lambda_0) = \int_{\Delta\lambda} \int_A E_\lambda(x, y, \lambda_0, \lambda) \cdot S_\Phi(x, y, \lambda) \cdot dA \cdot d\lambda \quad [U], \tag{3.1}$$

where $V(A, \Delta\lambda, \lambda_0)$ is the output signal (in U units, typically volts or amperes); $E_\lambda(x, y, \lambda_0, \lambda)$ is the spectral irradiance function in λ of the comparator system at detector position x,y for a wavelength setting of λ_0; $S_\Phi(x, y, \lambda)$ is the spectral (radiant flux) responsivity of the detector; A is the area of the radiant flux beam at the detector (See fig. 3.1.); and $\Delta\lambda$ is the wavelength interval for which the value of E_λ is not zero (i.e., the full-width bandpass). This equation is equivalent to eq (7.1) [26], with some minor changes in the notation, where A and $\Delta\lambda$ are left symbolically in V to indicate that V depends on how these elements are chosen.

To simplify the analysis, the responsivity S_Φ is assumed uniform throughout A, so,

$$S_\Phi(x, y, \lambda) = S_\Phi(\lambda) \quad [U \cdot W^{-1}]. \tag{3.2}$$

Thus, the responsivity S_Φ is no longer dependent on position (x,y) and can be removed from the area integral

$$V(A, \Delta\lambda, \lambda_0) = \int_{\Delta\lambda} S_\Phi(\lambda) \cdot \int_A E_\lambda(x, y, \lambda_0, \lambda) \cdot dA \cdot d\lambda \quad [U]. \tag{3.3}$$

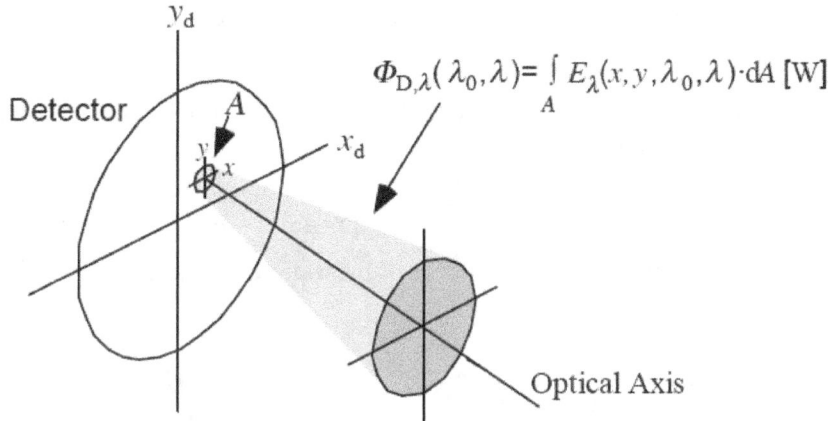

Figure 3.1. The geometry for detector spectral responsivity measurements.

Evaluating the area integral gives the spectral radiant flux function of the comparator system at the detector

$$\Phi_{D,\lambda}(\lambda_0,\lambda) = \int_A E_\lambda(x,y,\lambda_0,\lambda) \cdot dA = \tau(\lambda) \cdot \Phi_\lambda(\lambda_0,\lambda) \ [W], \quad (3.4)$$

where $\Phi_\lambda(\lambda_0,\lambda)$ is the output flux from the monochromator and $\tau(\lambda)$ is the transmittance of the optics (and atmosphere) between the monochromator and the detector (See fig. 3.2.).

Considering only the spectral dependence of the signal for any given A, the measurement equation can be written

$$V(\Delta\lambda,\lambda_0) = \int_{\Delta\lambda} S_\Phi(\lambda) \cdot \tau(\lambda) \cdot \Phi_\lambda(\lambda_0,\lambda) \cdot d\lambda \ [U]. \quad (3.5)$$

Equation (3.5) is the flux equivalent to eq (7.1b) [26]. Introducing the slit-scattering function $z(\lambda_0 - \lambda)$ to the measurement equation allows the spectral radiant flux function of the monochromator $\Phi_\lambda(\lambda_0,\lambda)$ to be written as the product of two functions (with appropriate normalization)

$$\Phi_\lambda(\lambda_0,\lambda) = z(\lambda_0 - \lambda) \cdot \Phi_{f,\lambda}(\lambda) \ [W], \quad (3.6)$$

where the slit-scattering function $z(\lambda_0 - \lambda)$ is dependent only on the difference between the wavelength setting of the monochromator and the wavelength of the flux and the "spectral flux" factor $\Phi_{f,\lambda}$ is dependent only on the wavelength of the flux. The factor $\Phi_{f,\lambda}$ is the spectral radiant flux at λ_0, $\Phi_\lambda(\lambda_0)$, and is equivalent to the responsivity factor r^f introduced in eq (7.12) [26]. Both the slit-scattering function $z(\lambda_0 - \lambda)$ and the factor $\Phi_{f,\lambda}$ can be determined experimentally, although the latter requires deconvolution from the measured output flux of the monochromator.

Thus the output signal V can now be written

$$V(\Delta\lambda,\lambda_0) = \int_{\Delta\lambda} S_\Phi(\lambda)\cdot\tau(\lambda)\cdot z(\lambda_0-\lambda)\cdot\Phi_{f,\lambda}(\lambda)\cdot d\lambda \;[U]. \tag{3.7}$$

Equation (3.7) is the flux equivalent to eq (7.13) [26] and is in a form that can be used for finding the spectral responsivity of a detector. Thus a measurement equation has been developed for a spectrally selective source instead of a spectrally selective detector (spectroradiometer). If the product $S_\Phi\cdot\tau\cdot\Phi_{f,\lambda}$ is approximately constant (or linear) over the spectral range for which $z(\lambda_0-\lambda)$ is significant (i.e., within $\Delta\lambda$) and z is approximately symmetrical with respect to λ_0, eq (3.7) can be written

$$V(\Delta\lambda,\lambda_0) \cong S_\Phi(\lambda_0)\cdot\tau(\lambda_0)\cdot\Phi_{f,\lambda}(\lambda_0)\cdot\int_{\Delta\lambda} z(\lambda_0-\lambda)\cdot d\lambda \;[U]. \tag{3.8}$$

Equation (3.8) is the equivalent flux version of eq (7.18) [26]. Implicit in this derivation is the assumption that the slit-scattering function $z(\lambda_0-\lambda)$ does not change with the wavelength setting λ_0 of the monochromator (that is, the dispersion is the same for all wavelengths) and thus $z(\lambda_0-\lambda)$ need only be measured once. Optical aberrations, scattering, and diffraction are also assumed to be the same whether a monochromatic beam is varied in λ over the monochromator bandpass, $\Delta\lambda$, or λ_0 of the monochromator is varied over a beam of fixed wavelength λ.

Equation (3.7) is the measurement equation for an ideal monochromator. A real monochromator system has spectrally scattered light (also known as stray light or out-of-band radiation and hereafter will be referred to as simply stray light) due to imperfections in the monochromator (and other optics). This is light from outside the spectral region $\Delta\lambda$ which is scattered into $\Delta\lambda$ and which contributes to the measured signal. Adding a stray light term, $V_{sl}(\Delta\lambda,\lambda_0)$, to eq (3.8) gives

$$V(\Delta\lambda,\lambda_0) \cong S_\Phi(\lambda_0)\cdot\tau(\lambda_0)\cdot\Phi_{f,\lambda}(\lambda_0)\cdot\int_{\Delta\lambda} z(\lambda_0-\lambda)\cdot d\lambda + V_{sl}(\Delta\lambda,\lambda_0)\;[U], \tag{3.9}$$

where
$$V_{sl}(\Delta\lambda,\lambda_0) \cong \int_{\lambda\neq\Delta\lambda} S_\Phi(\lambda)\cdot z(\lambda_0-\lambda)\cdot\tau(\lambda)\cdot\Phi_{f,\lambda}(\lambda)\cdot d\lambda \;[U]. \tag{3.10}$$

The V_{sl} term in eq (3.9) is typically small for radiometric measurement systems and is normally ignored in the "routine" measurement equation but is included in the uncertainty estimate calculations in section 7.1.2. Equation (3.9) separates the measurement equation into two parts, the first term represents the in-band signal and the second term represents the out-of-band signal as indicated by the limits on each integral.

The integral remaining in the first term can be evaluated and combined with the "spectral flux" factor $\Phi_{f,\lambda}$ to give an expression for the flux leaving the monochromator Φ'_λ,

$$\Phi'_\lambda(\lambda_0) = \Phi_{f,\lambda}(\lambda_0)\cdot\int_{\Delta\lambda} z(\lambda_0-\lambda)\cdot d\lambda \cong \Phi_\lambda(\lambda_0,\lambda) \;[W]. \tag{3.11}$$

Using eq (3.11) the measurement equation (eq (3.9)) can now be written in a form that is easily applied to the situation of measuring detector spectral responsivity

$$V(\Delta\lambda, \lambda_0) \cong S_\Phi(\lambda_0) \cdot \tau(\lambda_0) \cdot \Phi'_\lambda(\lambda_0) \quad [\text{U}]. \tag{3.12}$$

Before developing a "routine" measurement equation, it is important to reiterate the simplifying assumptions that were made to develop eq (3.12).

3.1.1 Approximations

The assumptions made in the development of the measurement equation (eq (3.12)) are summarized and discussed below. The development of the measurement equation followed the procedure described in chapter 5 of the Self-Study Manual on Optical Radiation Measurements [25] and does not include parameters for time, polarization, or incident angle; nor does it include environmental parameters such as ambient temperature or humidity, corrections for diffraction effects (departure from geometrical (ray) optics), and nonlinear responsivity.

A key requirement to detector-based radiometry is that detector responsivities are stable over time. In general the light from any monochromator system can be polarized, thus the effect of polarization on the responsivity of the detectors needs to be evaluated. Generally, this is not a problem for most photodetectors since the detectors are measured at normal incidence to the optical axis (and the detector surfaces are isotropic). The effect of the converging beam angle on the reflectance (and transmittance) from the detector surface (and window) is small compared to the variance of repeated measurements and is typically neglected. However this is not the case when filters are used with the detectors (e.g., photometers), especially interference filters, where the transmission is a strong function of the angle of incidence. Also, the detector area must be larger than the optical beam so that all of the optical radiation is collected by the detector (i.e., the detectors are underfilled). This also requires that the detector Field-of-View (FOV) be larger than the optical beam from the comparator system. The detector size and FOV of the typical 1 cm^2 photodiodes NIST provides meet these requirements.

Humidity can affect the measurement by changing the transmittance τ of the system and will be taken into account in the measurement equation developed below. Humidity can also affect the detector itself (or its window). For example, with windowless silicon photodiodes the absorption of water by the SiO$_2$ surface passivation layer changes the photodiode reflectivity which changes the responsivity [27]. Typically, for a windowed detector in the laboratory, effects due to water absorption (onto the detector or window) is not observed.

The temperature variation of the laboratory is small (typically < 1 °C) over the measurement time; therefore, the responsivity temperature dependence is neglected over most of the spectral region. When of concern, it can be applied as an additional uncertainty term.

Diffraction effects can be estimated from [28]

$$\theta^d = 2.44 \cdot \frac{\lambda}{d} \quad [\text{rad}], \tag{3.13}$$

where θ^d is the diffraction angle (for the first Airy disk), λ is the longest wavelength of the system, and d is the diameter of the aperture at the monochromator exit slit (the smallest aperture in the system). For the Vis/NIR SCF $\lambda \approx 2$ μm and $d \approx 1$ mm which gives

$$\theta^d_{Vis/NIR} = 2.44 \cdot \frac{2\ \mu m}{1\ mm} = 4.88 \times 10^{-3} \approx 5\ mrad. \qquad (3.14)$$

The Vis/NIR SCF has a focal ratio (*f*-number or *f*/#) of ≈9 giving a beam angle of ≈110 mrad which is over 20 times greater than the diffraction limit. For the UV SCF ($\lambda \approx 0.5\ \mu m$, $d \approx 1.5$ mm, and *f*/# ≈5) the diffraction effects are even smaller due to the shorter wavelengths. In this case, the beam angle is ≈200 times greater than the diffraction limit. Thus diffraction effects can be ignored for most UV, visible, and near-IR systems of the type described in this publication. Likewise, coherence effects are negligible [29] and are ignored in the analysis of these comparator systems.

The responsivity S_Φ is assumed uniform throughout the area A of the incident beam on the detector. In practice, this is accomplished by reducing A (both mechanically with apertures and optically via imaging optics) so that S_Φ is uniform over A. This limits the amount of flux (power) that can be delivered to the detector but this does not hinder the responsivity measurements for typical photodiodes. Deviations from this approximation are considered as uncertainty terms.

Certain assumptions were made about the system [26] (primarily the monochromator). One assumption is that the dispersion remains the same for all wavelengths and that the slit-scattering function $z(\lambda_0 - \lambda)$ does not change with the wavelength setting λ_0 of the monochromator. These approximations are not included in the uncertainties because their effect was determined to be negligible. This allows $z(\lambda_0 - \lambda)$ to be measured only once for an instrument and used over the entire spectral range. Another assumption is that optical aberrations, scattering, and diffraction are the same whether the λ of a monochromatic source is varied over the monochromator bandpass $\Delta\lambda$ or whether the monochromator wavelength setting λ_0 is varied over a monochromatic source of wavelength λ. Also $z(\lambda_0 - \lambda)$ is assumed approximately symmetrical with respect to λ_0 which is the case for the monochromators described in this publication.

The product $S_\Phi \cdot \tau \cdot \Phi_{f,\lambda}$ is assumed to be constant (or linear) over $\Delta\lambda$. In practice, this is affected by reducing $\Delta\lambda$ so that $S_\Phi \cdot \Phi_{f,\lambda}$ is constant (or linear) over $\Delta\lambda$. This also limits the amount of flux (power) that can be delivered to the detector, but this is not a hindrance for typical measurements. Most measurements are made on broadband detectors using a broadband source to provide a wide range of measurement wavelengths. However, this assumption may not be valid with some arc and discharge sources (primarily used in the UV) that can have strong spectral variations in output intensity. Typically, they are small and considered as uncertainty terms.

The spectrally scattered light is typically negligible for double monochromator systems but not for many single monochromator systems. Because the spectrally scattered light is small in magnitude it will not be used as a correction term in the measurement equation, but it will be evaluated and considered as an uncertainty term in section 7.

3.2 Substitution Method

The substitution method uses a standard detector to transfer its responsivity (output divided by input) to an unknown (test) detector. This method has a number of advantages that will be discussed in general and applied specifically to photodetectors. The measurement equation will then be developed and the general assumptions and their consequences discussed. Finally the measurement equation as applied to the systems described in this publication is developed.

3.2.1 General Substitution Method

The output (signal) Y from a detector (or system) with a linear response in U units, typically volts or amperes, can be written in general

$$Y_x = S_x \cdot X_x \text{ [U]}, \tag{3.15}$$

where Y_x is the signal from the detector, S_x is the responsivity of the detector, and X_x is the input to the detector. The signal from a standard detector with a known responsivity S_s is given by

$$Y_s = S_s \cdot X_s \text{ [U]}. \tag{3.16}$$

Dividing eq (3.15) by eq (3.16) we have

$$\frac{Y_x}{Y_s} = \frac{S_x \cdot X_x}{S_s \cdot X_s}. \tag{3.17}$$

If the input X to each system is assumed to be constant (this is the basis of the substitution method), solving for the detector responsivity S_x gives

$$S_x = \frac{Y_x}{Y_s} \cdot S_s \text{ [U·input units}^{-1}\text{]}. \tag{3.18}$$

Thus, using the substitution method, the standard detector responsivity S_s is scaled by the ratio of the outputs. The transfer is accomplished in essence by the standard detector measuring the input X to the test detector.

3.2.2 Photodetector Substitution

Photodetector responsivity measurements by detector substitution can be made using eq (3.18) and the following equipment: a broadband source, monochromator, focusing optics, and standard detector. (See fig. 3.2.) The radiant flux (power) Φ is the output flux from the source that enters the monochromator. The spectral radiant flux Φ_λ is the output flux from the monochromator and τ is the transmittance of any optics (and the atmosphere) between the monochromator and the detector. The spectral radiant flux received by the detectors is $\Phi_{D,\lambda} = \tau \cdot \Phi_\lambda$.

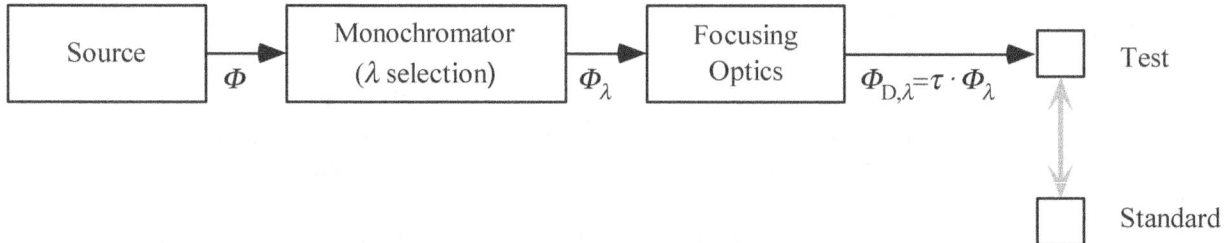

Figure 3.2. Block diagram of photodetector substitution method.

Using eq (3.18) assumes that the source is stable over the comparison time. This is generally true for incandescent sources (i.e., quartz-halogen (QTH) lamps), but less true for arc sources which are primarily used in the UV.

3.2.3 Substitution Method with Monitor

To eliminate the effect of power fluctuations, a beamsplitter and monitor detector are used. (See fig. 3.3.) Here $\Phi_{D,\lambda} = \tau_{bs} \cdot \tau \cdot \Phi_\lambda$ is the flux received by the detectors and $\Phi_{M,\lambda} = \rho_{bs} \cdot \tau \cdot \Phi_\lambda$ is the flux received by the monitor. The transmittance and reflectance of the beam splitter are τ_{bs} and ρ_{bs}, respectively. The monitor detector and beamsplitter are assumed to be stable (constant) over the measurement time. Thus the monitor records the source power fluctuations and the ratio of detector to monitor signals will be constant.

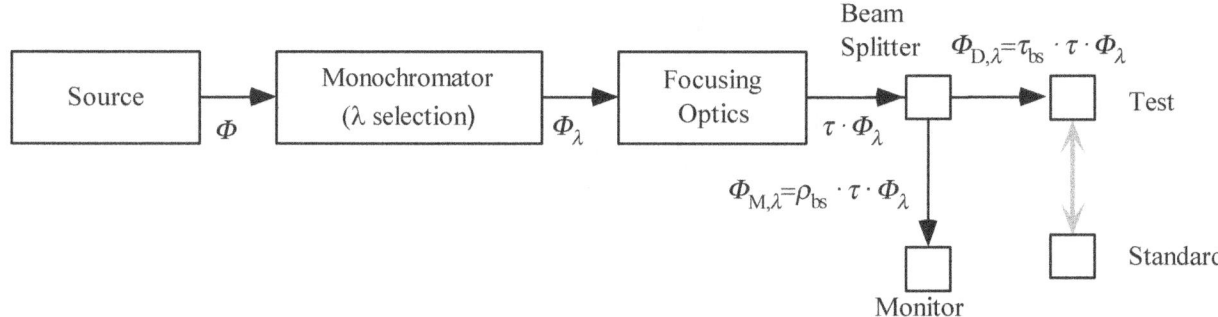

Figure 3.3. Block diagram of photodetector substitution method with monitor.

3.2.4 Measurement Equation Applied to the SCFs

The actual measurement equation is now developed. Dropping the λ notation and changing the sub- and superscripts for clarity from eq (3.12), it can easily be shown that the signal from a detector x is

$$V_x = S_x \cdot G_x \cdot \tau_x \cdot \Phi_x + V_{d,x} \; [V], \qquad (3.19)$$

where S_x is the spectral responsivity in $A \cdot W^{-1}$, G_x is an explicit gain term for a transimpedance amplifier in $V \cdot A^{-1}$, τ_x is the transmittance of any optics (and the atmosphere) between the monochromator and the detector, Φ_x is the output flux from the monochromator in W, and $V_{d,x}$ is the dark output (sometimes called the "background signal") in V, i.e., the signal produced when no flux is incident on the detector. (For photodiodes this is caused by their dark current.) In practice $V_{d,x}$ is found by measuring the signal from the detector with the shutter closed at the exit slit of the monochromator. The net signal ΔV_x is the signal due to the radiant (optical) flux received by the detector

$$\Delta V_x = V_x - V_{d,x} = S_x \cdot G_x \cdot \tau_x \cdot \Phi_x \ [\text{V}]. \tag{3.20}$$

This is the "routine" measurement equation for the substitution method depicted in figure 3.2. The measurement equation for the "substitution method with monitor" (fig. 3.3) is written by explicitly including the transmittance τ_{bs} and reflectance ρ_{bs} of the beam splitter assembly (which also contains the reflectance of the monitor turning mirror). The systems described in this publication simultaneously sample both the detector and monitor. The signals from the test detector x and monitor detector mx are

$$\Delta V_x = V_x - V_{d,x} = S_x \cdot G_x \cdot \tau_{bsx} \cdot \tau_x \cdot \Phi_x \ [\text{U}], \tag{3.21}$$

and
$$\Delta V_{mx} = V_{mx} - V_{d,mx} = S_{mx} \cdot G_{mx} \cdot \rho_{bsx} \cdot \tau_x \cdot \Phi_x \ [\text{U}]. \tag{3.22}$$

The ratio of these two signals is

$$R_x = \frac{\Delta V_x}{\Delta V_{mx}} = \frac{S_x \cdot G_x \cdot \tau_{bsx} \cdot \tau_x \cdot \Phi_x}{S_{mx} \cdot G_{mx} \cdot \rho_{bsx} \cdot \tau_x \cdot \Phi_x}, \tag{3.23}$$

where $\tau_{bsx} \cdot \tau_x \cdot \Phi_x$ is the fraction of light received by the test detector, and $\rho_{bsx} \cdot \tau_x \cdot \Phi_x$ is the fraction of light received by the monitor.

There is a similar ratio for the two signals from the working standard detector and monitor,

$$R_s = \frac{\Delta V_s}{\Delta V_{ms}} = \frac{S_s \cdot G_s \cdot \tau_{bss} \cdot \tau_s \cdot \Phi_s}{S_{ms} \cdot G_{ms} \cdot \rho_{bss} \cdot \tau_s \cdot \Phi_s}. \tag{3.24}$$

Dividing eq (3.23) by eq (3.24) (taking the ratio of the ratios) gives

$$\frac{R_x}{R_s} = \frac{S_x \cdot G_x \cdot \tau_{bsx} \cdot \tau_x \cdot \Phi_x}{S_{mx} \cdot G_{mx} \cdot \rho_{bsx} \cdot \tau_x \cdot \Phi_x} \cdot \frac{S_{ms} \cdot G_{ms} \cdot \rho_{bss} \cdot \tau_s \cdot \Phi_s}{S_s \cdot G_s \cdot \tau_{bss} \cdot \tau_s \cdot \Phi_s}. \tag{3.25}$$

Looking at eq (3.25), it is now seen that variations or drifts in the source flux or system transmittance during the time between the measurement of the test detector and the working standard detector are canceled by the monitor detector. If the beamsplitter and monitor detector responsivity and amplifier gain are stable (constant) over the comparison time, other terms cancel, leaving

$$\frac{R_x}{R_s} = \frac{S_x \cdot G_x}{S_s \cdot G_s}. \tag{3.26}$$

Solving eq (3.26) for the test detector spectral responsivity S_x and substituting the signal measurements into the ratios we have

$$S_x = \frac{R_x}{R_s} \cdot \frac{G_s}{G_x} \cdot S_s = \frac{\dfrac{V_x - V_{d,x}}{V_{mx} - V_{d,mx}}}{\dfrac{V_s - V_{d,s}}{V_{ms} - V_{d,ms}}} \cdot \frac{G_s}{G_x} \cdot S_s \ [\text{A·W}^{-1}]. \qquad (3.27)$$

This is the working form of the measurement equation. Note that it still has the general form

$$S_x = \frac{Y_x}{Y_s} \cdot S_s \ [\text{U·input units}^{-1}]. \qquad (3.18)$$

Depending on how the monochromator wavelength scale is calibrated, a correction term may be needed when the responsivity curves of the test and working standard detectors have different slopes. If the centroid wavelength of the bandpass is used to calibrate the monochromator wavelength scale, then there is no correction term. But if the peak wavelength of the bandpass is used, then eq (3.27) requires a correction term. Thus the measurement equation becomes

$$S_x = \frac{R_x}{R_s} \cdot \frac{G_s}{G_x} \cdot S_s + C_{bw} \ [\text{A·W}^{-1}], \qquad (3.28)$$

where C_{bw} is a correction term due to the bandpass of the monochromator and is referred to as the "bandwidth-effect." Like the stray light term, the bandwidth-effect is small for this system and is ignored in the "routine" measurement equation. The bandwidth-effect is analyzed in section 7.1.2 as an uncertainty term and included in the uncertainty estimate.

4. Equipment Description

In this section, the Visible to Near-Infrared Spectral Comparator Facility (Vis/NIR SCF) and Ultraviolet Spectral Comparator Facility (UV SCF) components are described along with the associated electronics.

4.1 Visible to Near-Infrared (Vis/NIR) Comparator Description

The Visible to Near-Infrared Spectral Comparator Facility (Vis/NIR SCF) is a monochromator-based system that typically measures the uniformity and absolute spectral responsivity of photodiodes in the 350 nm to 1800 nm spectral region. The Vis/NIR SCF operates from 350 nm to 1100 nm using silicon photodiodes as working standards and from 700 nm to 1800 nm using germanium photodiodes as working standards.

The Vis/NIR SCF uses the direct substitution method and automated translation stages to position the photodetectors for measurement. The test detectors as well as the working standards are fixed onto optical mounts that rotate and tilt for accurate alignment. A variety of sources can be selected. Typically a 100 W quartz-halogen lamp is used as the source in the Vis/NIR SCF. A shutter is located just after the monochromator exit slit. A monitor detector located after the monochromator measures source fluctuations. The detectors and the exit optics are enclosed in a light tight box. A diagram of the Vis/NIR SCF is shown in figure 4.1. Each of the primary components is now described in greater detail.

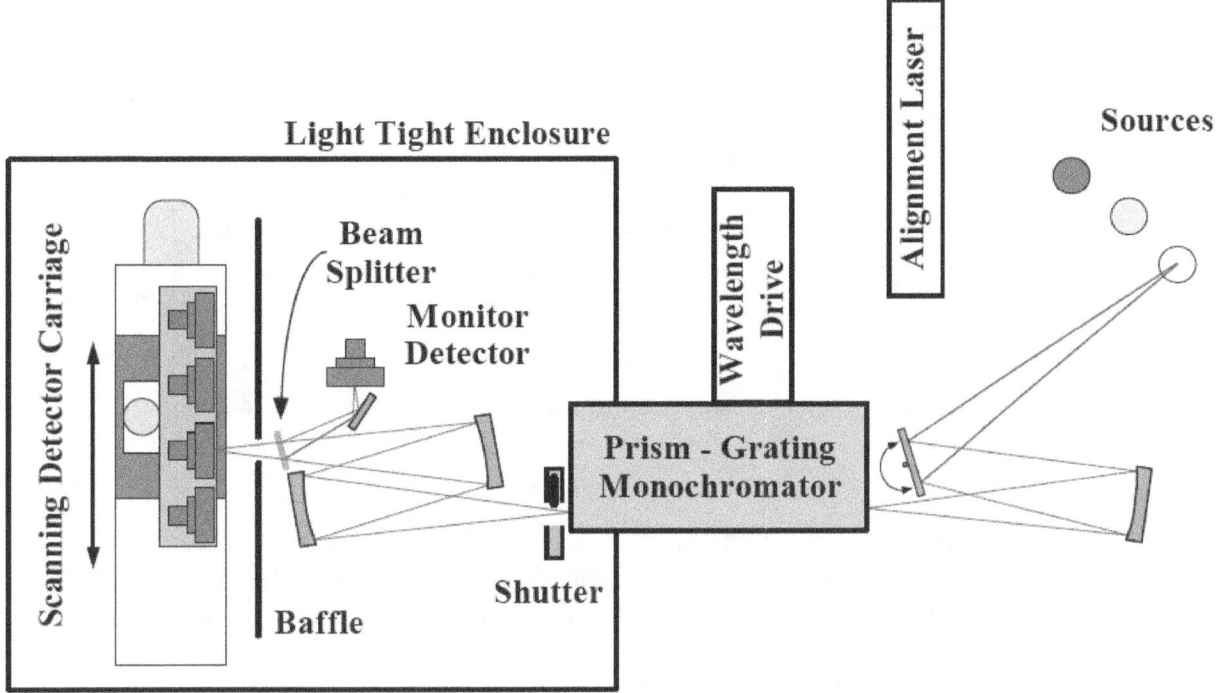

Figure 4.1. Visible to Near-Infrared Spectral Comparator Facility (Vis/NIR SCF).

4.1.1 Vis/NIR Source

Typically the Vis/NIR SCF uses a 100 W (12 V) quartz-halogen lamp that is constant-voltage controlled for stability. The lamp is used over the spectral range of 350 nm to 1800 nm. An FEL lamp, tungsten strip lamp, xenon arc lamp, argon arc, or alignment laser can be used as a source in the Vis/NIR comparator as well. The spectral output flux (at the detector) of the 100 W quartz-halogen lamp with the Vis/NIR SCF is shown in figure 4.2.

Safety considerations

Quartz-halogen lamps, lasers, and arc sources are potential eye hazards. Care is taken to never look directly into any of the sources or the laser beam. Shields mounted on the side of the optical table inhibit direct visual contact with the sources used with the comparator. The lasers used for alignment are Laser Safety Class II. Protective eyeware is worn when working in proximity to these sources.

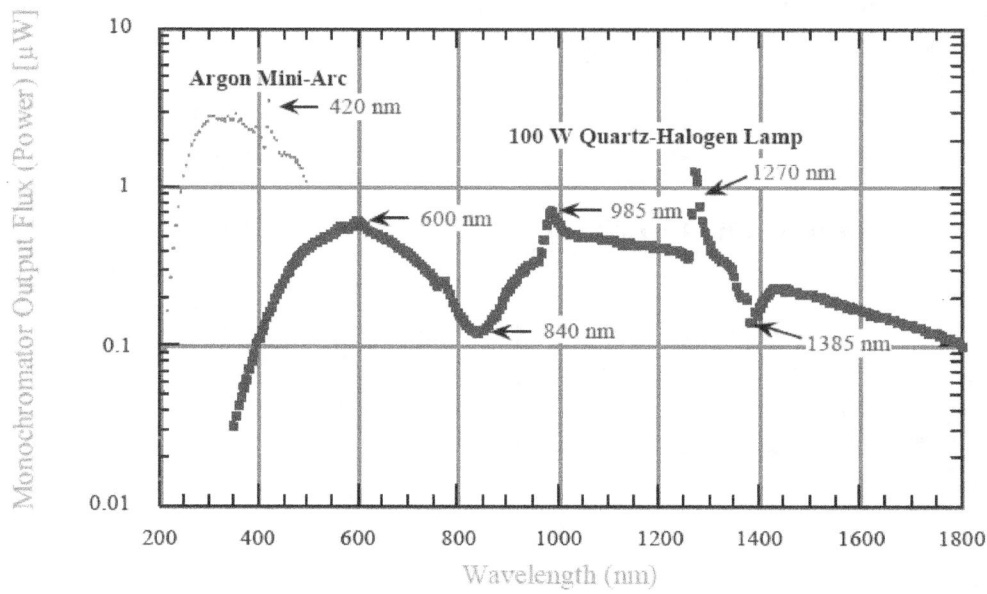

Figure 4.2. Spectral output flux of the UV and visible to near-IR monochromators. The argon mini-arc data are from the UV monochromator and 100 W quartz-halogen lamp data are from the Vis/NIR monochromator.

4.1.2 Vis/NIR Monochromator

The visible comparator uses a prism-grating monochromator, a NIST modified Cary-14, which employs a 30° fused silica prism in series with a 600 lines per mm echelette grating. The monochromator's spectral range is 186 nm to 2.65 µm. The spectral range used in the detector characterization facility is 350 nm to 1800 nm. In the typical measurement configuration, the monochromator slits are set to 1.1 mm with a bandpass of 4 nm. A circular aperture of 1.1 mm diameter just past the exit slit determines the beam size. The exit beam is $f/9$. The monochromator has a stray light rejection of 10^{-8}. Greater than 99 % of the beam flux lies within an area of 1.6 mm diameter around the optical axis.

4.1.3 Vis/NIR Optics

A 46 mm x 61 mm flat mirror rotates on an automated stage for source selection. The source is imaged by a stationary 15.24 cm diameter spherical mirror onto the entrance slit of the monochromator. A shutter, with a 25 mm diameter aperture, is placed inside the enclosure past the monochromator exit slit and aperture.

Two 15.24 cm diameter spherical mirrors and two 7.62 cm diameter flat mirrors image the exit aperture of the monochromator with 1:1 magnification onto the detector. The two flat mirrors used to optically fold the system are not shown in figure 4.1. Mirrors are used to prevent chromatic aberrations, and the astigmatism of the spherical mirrors is corrected (to first order) by tilting the second spherical mirror perpendicular to the plane of the first spherical mirror [30].

4.1.4 Vis/NIR Translation Stages - Detector Positioning

The Vis/NIR comparator uses two orthogonal linear positioning stages to translate the test detectors and working standard detectors. The horizontal stage travel range is 400 mm with a manufacturer specified resolution of 0.1 μm and an accuracy of 0.2 μm per 100 mm. The vertical stage travel range is 50 mm with a manufacturer specified resolution of 0.1 μm and an accuracy of 0.25 μm per 25 mm.

Each detector can be translated along the optical axis for focusing. A gimbal mount allows the rotation and tilt of each detector to be adjusted perpendicular to the optical axis.

4.1.5 Vis/NIR Working Standards

The visible working standards (Vis WS) are four Hamamatsu S1337-1010BQ silicon photodiodes with fused quartz windows. This is a p-n photodiode with a 1 cm^2 active area. The Hamamatsu S1337 diode is a popular diode for radiometric standards, and the linearity, uniformity, and stability of the Hamamatsu S1337 diode has been documented in many places [31, 32]. The Vis WS are normally used for 350 nm to 1100 nm measurements. For near-infrared measurements (700 nm to 1800 nm), four EG&G Judson J16TE2-8A6-R05M-SC temperature-controlled germanium photodiodes are used as working standards (Ge WS). This is a p-n photodiode with a 5 mm diameter active area and a wedged sapphire window. Typically, two working standards are used for each spectral comparison measurement. The second working standard is used as a "check" standard.

4.1.6 Beam Splitter and Monitor Detector

Variations in the source intensity are corrected by using a beam splitter and monitor detector as discussed in section 3.2.3. The beam splitter is a 50.8 mm x 50.8 mm flat fused quartz plate. Hamamatsu S1337-1010BQ and EG&G Judson J16TE2-8A6-R05M-SC photodiodes are used as monitor detectors. The monitor detector is typically the same type as the working standard used for the measurement. A 35 mm x 46 mm flat oval mirror is used to optically fold the monitor beam to a convenient physical location for the monitor.

4.1.7 Alignment Lasers

Two HeNe lasers are aligned to the optical axis of the Vis/NIR comparator defined by the monochromator. One laser is located in one of the source positions and is typically used to align the detectors. The other is located inside the enclosure and is pointed "backwards" through the comparator to align the sources. It is not shown in figure 4.1.

4.1.8 Enclosure

Each comparator has a light tight enclosure to eliminate the background light from the room. This is essential for the routine dc (optically unmodulated) measurements. The enclosure is a box that sits on the optical table. Several doors allow easy access to the equipment and detectors. The enclosure houses the test detectors, working standards, monitor detector, and associated electronics, including the amplifiers. The horizontal and vertical translation stages and the exit optics are housed inside the enclosure as well.

The enclosure also reduces the amount of dust settling on the detectors and exit optics. An exhaust fan is used to regulate the environment inside of the enclosure, and the temperature inside the enclosure is monitored.

4.2 Ultraviolet (UV) Comparator Description

The Ultraviolet Spectral Comparator Facility (UV SCF) is a monochromator-based system that measures the uniformity and absolute spectral responsivity of photodetectors in the 200 nm to 500 nm spectral region. The UV SCF is very similar in configuration and operation to the Vis/NIR SCF. Only the differences between the two will be described. A diagram of the UV SCF is shown in figure 4.3.

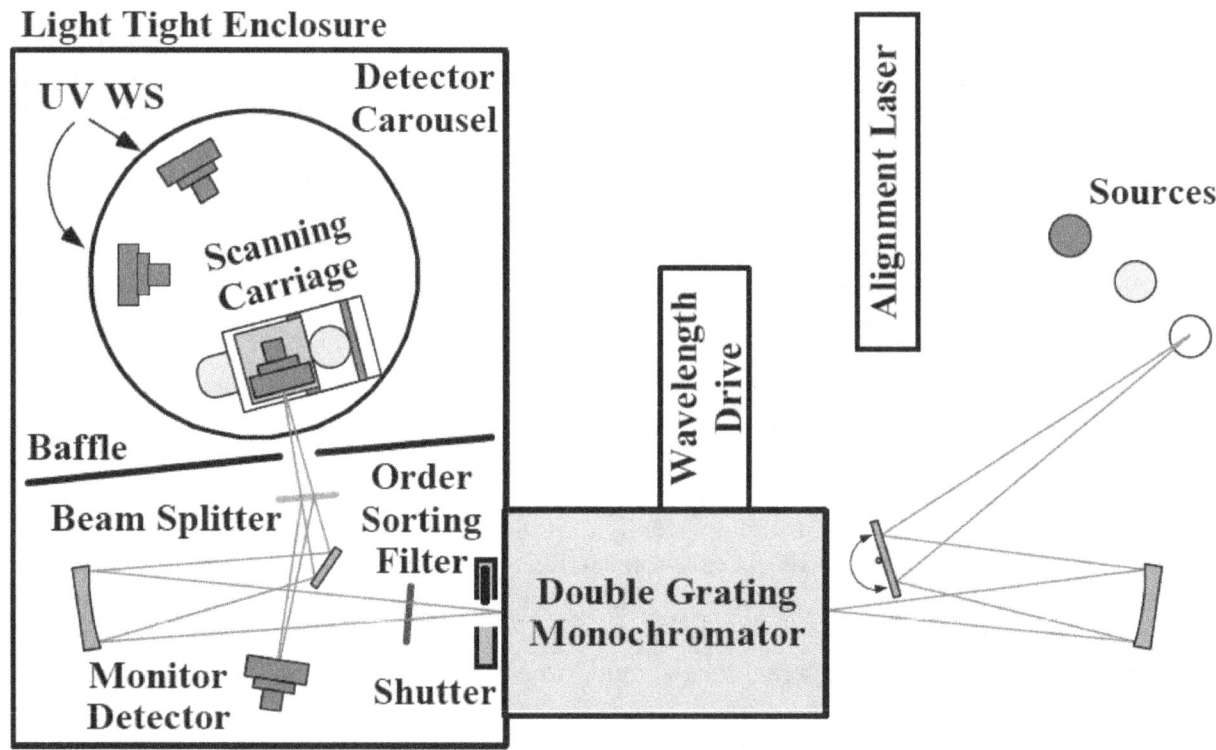

Figure 4.3. Ultraviolet Spectral Comparator Facility (UV SCF).

UV enhanced silicon photodiodes serve as the working standards for the UV SCF. A rotary stage is used in the UV SCF and currently only one test detector at a time can be measured. The test and working standard detectors are fixed onto optical mounts that rotate and tilt. But only the test detector can automatically be positioned in the horizontal and vertical planes. The working standards are positioned manually. Typically an argon arc is used as the source in the UV SCF.

4.2.1 UV Source

The UV SCF primarily uses a NIST-designed argon mini-arc as its source over the 200 nm to 500 nm spectral range. The argon mini-arc was developed at NIST as a secondary spectral radiance standard and has been well characterized [33]. The argon mini-arc is an intense, uniform uv source, with argon gas flowing through the arc structure at $\approx$ 35 kPa. The arc can be operated from 30 A to 100 A, but is typically operated at $\approx$ 40 A (and 60 V). The arc is cooled using the normal facility 5.6 °C (42 °F) chilled water. The water runs through plastic nonconducting tubing. An FEL lamp, tungsten strip lamp, xenon arc lamp, 100W quartz-halogen lamp, or alignment laser can be used as a source in the UV comparator as well. The spectral output flux (at the detector) of the argon mini-arc with the UV SCF is shown in figure 4.2.

Safety considerations

UV filtering safety glasses are worn whenever the argon mini-arc source is lit. Even with safety glasses on, the argon mini-arc is never viewed directly. Shields mounted on the side of the optical table inhibit direct visual contact with the sources. Since very high currents are used, extra precautions are taken when using the arc. Care is taken to avoid accidental contact with the arc electrical contacts and to insulate the arc from the optical table (and surroundings). Also, to avoid overheating the arc and electrical shorts, attention is paid that the cooling water is flowing before the arc is turned on and that no water leaks exist. Nonconducting plumbing is always used. The alignment lasers are Laser Safety Class II.

4.2.2 UV Monochromator

The UV comparator uses a Spex 1680, 1/4 m, double grating monochromator utilizing 2840 lines per mm gratings. The monochromator's spectral range is 180 nm to 1000 nm. The spectral range used in the detector comparator facility is 200 nm to 500 nm. In the typical measurement configuration, the entrance and exit slits are circular 1.5 mm diameter apertures with a bandpass of 4 nm. The exit beam is $f/5$. Greater than 99 % of the beam flux lies within an oval area of diameters 2.0 mm and 2.5 mm around the optical axis.

4.2.3 UV Optics

A 10.16 cm diameter flat mirror on a rotary stage is used for source selection. The source is imaged by a stationary 15.24 cm diameter spherical mirror onto the entrance slit of the monochromator. A shutter, with a 25 mm diameter aperture, is placed after the monochromator exit slit inside the enclosure.

One 15.24 cm diameter spherical mirror and one 7.62 cm (3 in) diameter flat mirror images the exit aperture of the monochromator with 1:1 magnification. Mirrors are used to prevent chromatic aberrations.

4.2.4 UV Translation Stages - Detector Positioning

The UV comparator has optical mounts for one test detector and two working standard detectors on a rotary positioning stage. The maximum travel of the rotary stage is 360° with a manufacturer specified resolution of 0.001° and an accuracy of 0.0014°. The test detector is mounted on two automated orthogonal linear translation stages. The stages have a travel range of 50 mm with a manufacturer specified resolution of 0.1 μm and an accuracy of 0.25 μm per 25 mm.

The working standard detectors are mounted on two manual orthogonal linear translation stages. Each detector is mounted to allow its position to be adjusted along the optical axis (focus). A gimbal mount allows the rotation and tilt of each detector to be adjusted perpendicular to the optical axis.

4.2.5 UV Working Standards

The ultraviolet working standards (UV WS) are four UDT Sensors UV100 silicon photodiodes with quartz windows and 1 cm^2 circular active areas. The UV100 is an inverted channel diode with enhanced resistance to UV radiation damage. Typically two working standards are used for each spectral comparison measurement.

4.2.6 Beam Splitter and Monitor Detector

The beam splitter is a 50.8 mm diameter flat quartz plate. A UDT Sensors UV100 photodiode is the monitor detector.

4.2.7 Alignment Lasers

Similar to the Vis/NIR SCF, two HeNe lasers are used to align the optical path of the UV comparator. The second laser is not shown in figure 4.3.

4.2.8 Enclosure

The UV SCF enclosure is similar in design and construction to that of the Vis/NIR SCF.

4.3 Electronics

4.3.1 Electronics - Signal Measurement

This section describes the electronics used with the UV SCF and Vis/NIR SCF for detecting and amplifying the signals from the photodetectors. The electronics for both comparators are identical with some equipment shared between the two facilities. Only the Vis/NIR SCF will be described and the differences between the comparators will be noted.

Figure 4.4 shows a block diagram of the typical setup for measurements in the Vis/NIR SCF. The design of the electronics and control of the UV SCF is very similar to the Vis/NIR SCF. Four NIST built and characterized amplifiers are housed in one module for convenience and a separate single amplifier is used with the monitor detector. The digital voltmeters (DVMs) are computer controlled via an IEEE-488 bus. The two DVMs simultaneously measure the signals from the monitor and one of the four detectors that can be moved into the SCF beam. One DVM has a multiplexed input for selecting any of four amplifier channels or temperature monitoring inputs. (Some detectors have temperature monitoring circuitry that produces a voltage signal proportional to their temperature.)

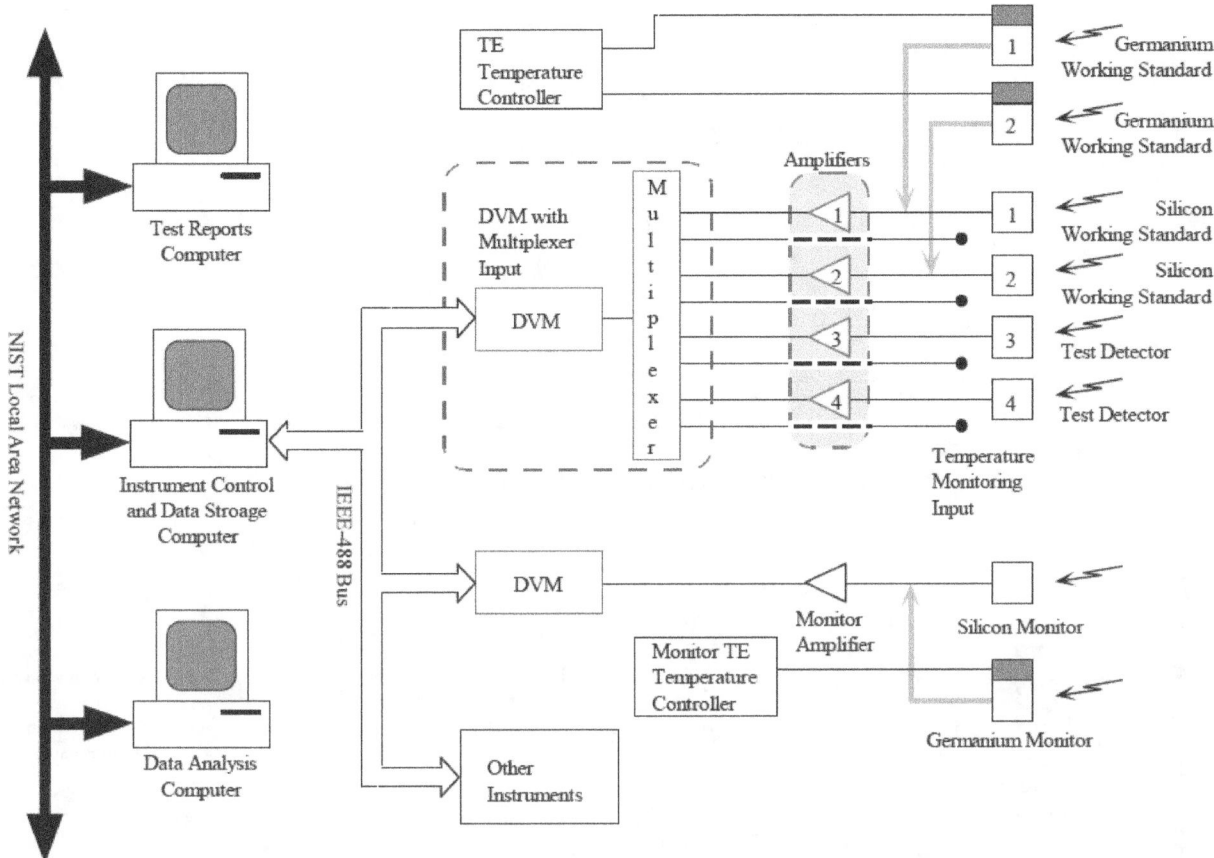

Figure 4.4. Vis/NIR detector, amplifier, DVM, and computer control block diagram.

The amplifiers are identical in design and operate with gains of 10^4 V·A^{-1} to 10^9 V·A^{-1}. References [34 and 35] and several titles in the bibliography describe similar transimpedance amplifiers and their operation. The schematic for the precision transimpedance (I/V) amplifiers used in both the UV and Vis/NIR SCFs is shown in figure 4.5.

Figure 4.6 shows a block diagram for chopped (ac) measurements. Four additional pieces of equipment are required: two lock-in amplifiers, an optical chopper, and a signal multiplexer. In this configuration, the outputs of the four amplifiers from the detectors are fed through the multiplexer and then to a lock-in amplifier. The monitor signal is fed straight from the amplifier to the lock-in. Both lock-ins are run in a mode where the phase (which remains constant) is set to zero, thus the only output signal is the signal magnitude. This signal is sent to the appropriate DVM. The DVM input is multiplexed as before with the temperature monitoring inputs. A discussion of the frequency dependent gain characteristics for similar photodiodes and amplifiers is found in Ref. [36].

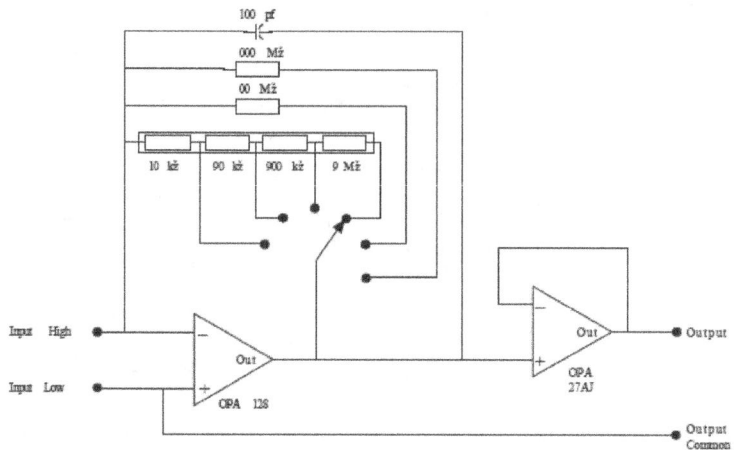

Figure 4.5. NIST SCF precision transimpedance (I/V) amplifier circuit.

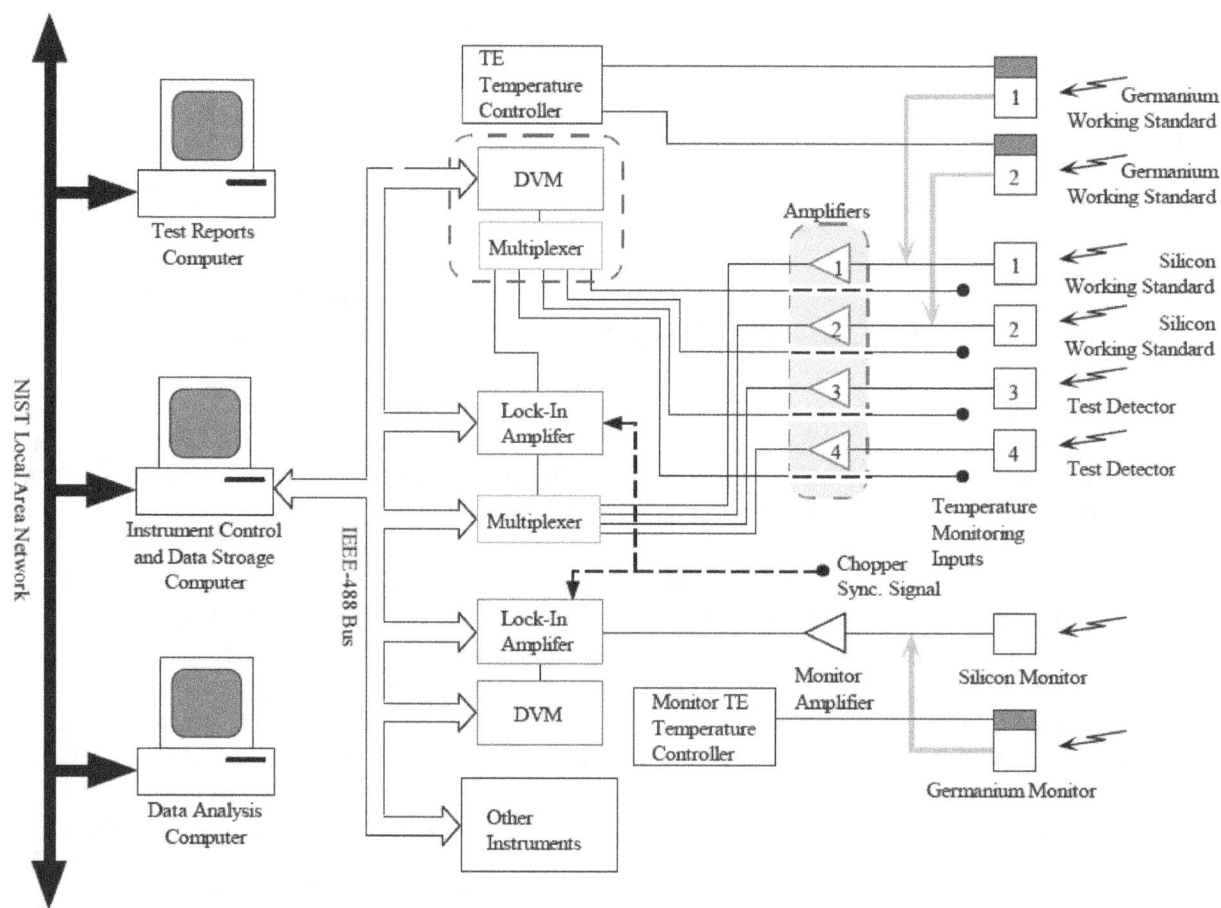

Figure 4.6. Vis/NIR detector, amplifier, lock-in, DVM, and computer control block diagram.

4.3.2 Electronics - Auxiliary Equipment

In both configurations, temperature-controlled germanium photodiodes can be used which require the use of the thermoelectric (TE) temperature controllers shown in figures 4.4 and 4.6. The following electronics not previously mentioned are also used routinely: a 4-channel TE temperature controller for the germanium working standards; a TE temperature controller for the germanium monitor; a digital I/O control unit for the optical shutter; and laboratory environmental monitoring system. The environmental monitoring system measures the temperature in each comparator enclosure, the laboratory temperature, relative humidity, barometric pressure, and the electrical power line voltage and frequency.

5. Absolute Spectral Responsivity Scale Realization

This section describes how the NIST absolute spectral responsivity scale and its traceability to the HACR are determined. First, the operation of the HACR and the scale transfer to the trap detectors are briefly reviewed. Then the transfer of the scale from the trap detectors to the visible working standards is described. Finally, the extension of the scale to the UV and near-IR working standards using a pyroelectric detector is explained. The scale realizations are planned annually.

5.1 Transfer from HACR to Traps (405 nm to 920 nm)

The HACR was constructed to improve the accuracy and spectral range of optical power (flux) measurements and is the U.S. primary standard for optical power (fig. 5.1). Cryogenic radiometers are currently used as the primary standard for optical power at other national laboratories [31, 37, 38]. The HACR will be briefly reviewed here. A full description can be found in Ref. [20]. The HACR is an electrical substitution radiometer (ESR) that operates by comparing the temperature rise induced by optical power absorbed in a cavity to the electrical power needed to cause the same temperature rise by resistive (ohmic) heating. Thus the measurement of optical power is determined in terms of the electrical watt in the form of voltage and resistance standards maintained by NIST [39]. Several advantages are realized by operating at cryogenic temperatures (≈ 5 K) instead of room temperature. The heat capacity of copper is reduced by a factor of 1000, thus allowing the use of a relatively large cavity. Also the thermal radiation emitted by the cavity or absorbed from the surroundings is reduced by a factor of $\approx 10^7$, which eliminates radiative effects on the equilibrium temperature of the cavity. Finally, the cryogenic temperature allows the use of superconducting wires to the heater, thereby removing the non-equivalence of optical and electrical heating resulting from heat dissipated in the wires. The relative combined standard uncertainty of the NIST HACR measurements is 0.021 % [21] at ≈ 1 mW. The largest components of the uncertainty are those due to the systematic correction for the Brewster angle window transmission and the random error associated with the cavity temperature measurement.

There are drawbacks for making routine measurements of photodetectors with the HACR. Because of its design, the only source that can be used with the HACR is a laser. Equipment setup and measurements with the HACR are very time consuming, typically taking several days for each wavelength. The measurement wavelengths are limited to available laser wavelengths. The power levels used for the highest accuracy measurements with the HACR are ≈ 0.8 mW which is higher than those desired for typical radiometric applications. Therefore a practical means of disseminating the optical power scale to customers is to transfer the scale to other NIST facilities specifically designed for transferring the optical power scale to customers (e.g., UV and Vis/NIR SCFs). Trap detectors make excellent transfer standards since they are stable, have uniform responsivity, good linearity, and low noise [21, 15, 31]. Figure 5.2 shows the arrangement of the photodiodes in the trap detectors used for HACR transfer standard detectors. Various detector arrangements acting as "light traps" for the light reflected from the first photodiode have been generically called "trap detectors." The light beam reflection into orthogonal planes reduces effects of polarization on the responsivity.

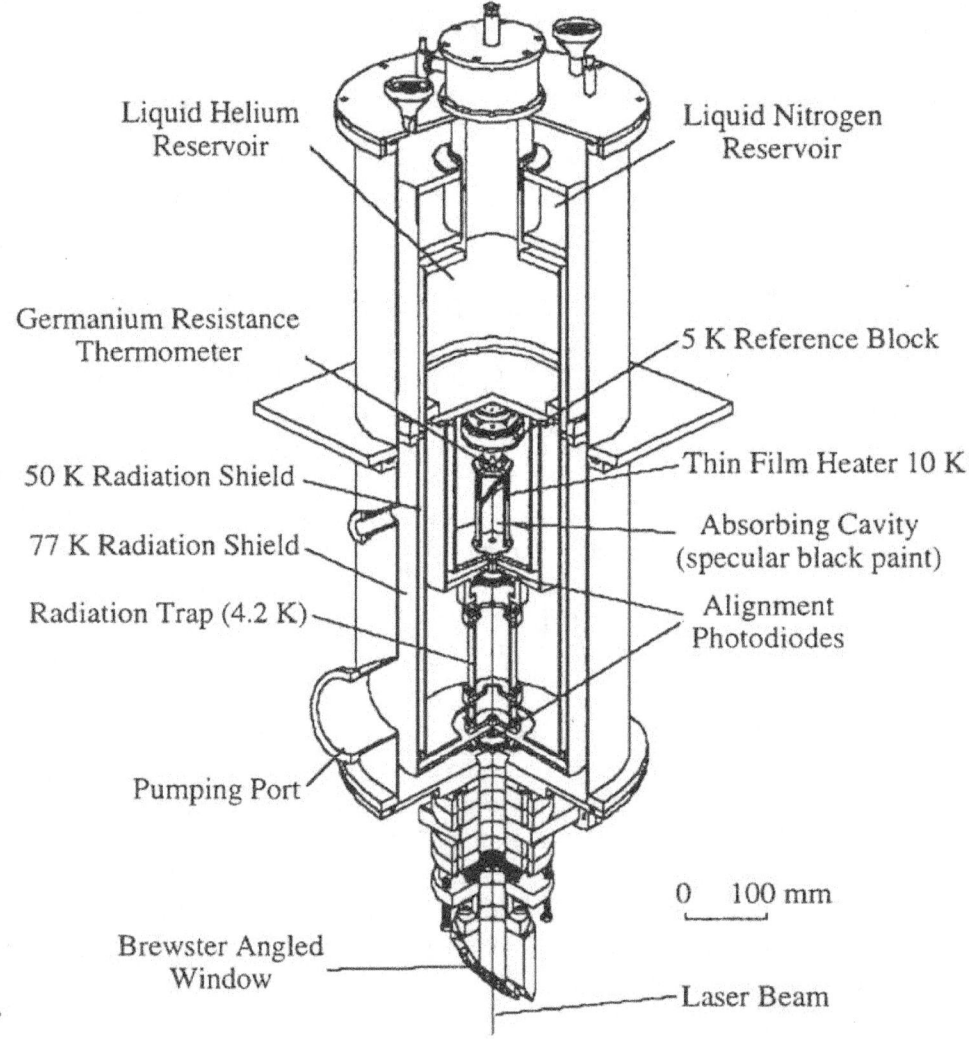

Figure 5.1. NIST High Accuracy Cryogenic Radiometer (HACR).

As a step in transferring the detector spectral power scale, the HACR was used to determine the external quantum efficiency (EQE) of two trap detectors in the visible spectral region (see fig. 5.3a). Figure 5.3b shows a diagram of the setup used to transfer the scale from the HACR to a trap detector using the substitution method. These trap detectors were calibrated at nine laser wavelengths between 406 nm and 920 nm. The reflectance of each trap detector was measured in this spectral range, allowing the internal quantum efficiency (IQE) to be determined. The IQE was modeled and the EQE for the entire spectral range was determined with a relative combined standard uncertainty of 0.03 %. Figure 5.3a shows the measured and modeled EQE of a trap detector. The trap detectors used in the scale transfer showed a maximum relative difference of 0.04 % for orthogonal linear polarizations [21]. Because the output from both of the SCFs is a mixture of polarizations (as with most monochromators) and different from the linear polarization of the lasers used with the HACR, this value is combined by the root-sum-of-squares method (explained in sec. 7.) with the relative combined standard uncertainty. The calibration relative standard uncertainty for the trap detectors when used with the SCFs is 0.05 %.

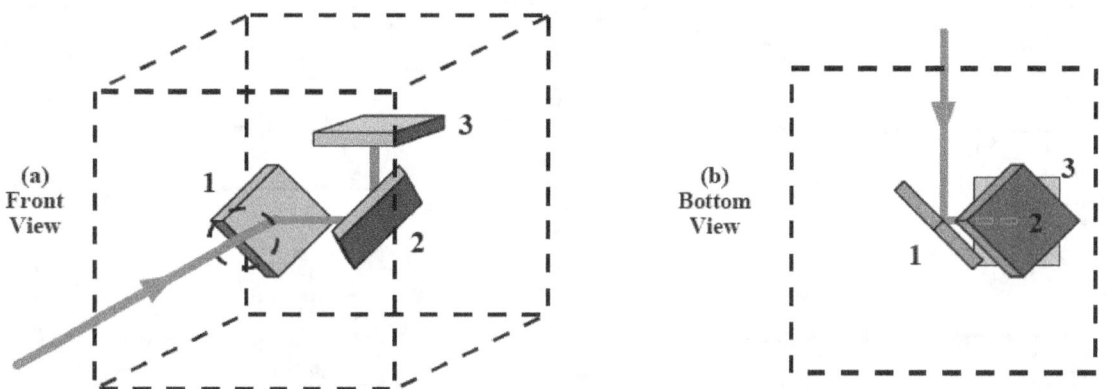

Figure 5.2. Trap detector arrangement of photodiodes minimizes light lost to reflections.

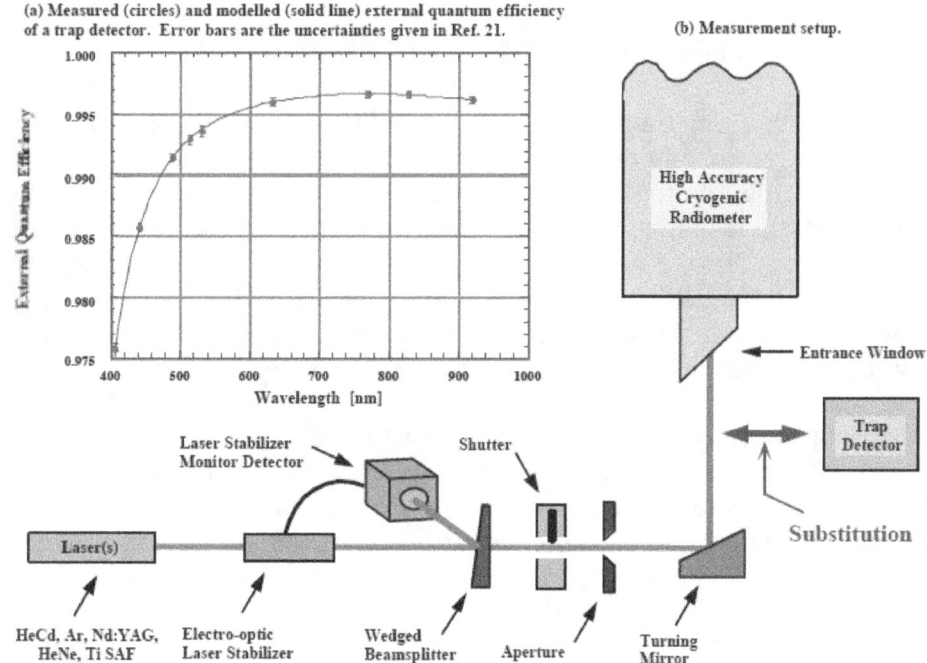

Figure 5.3. Scale transfer by substitution method with the HACR.

Figure 5.4 shows the entire spectral power scale propagation schematically. The HACR is the primary standard for the scale and trap detectors are used as transfer standards as described above. The trap detectors can then be used in a variety of measurements. First, the trap detectors could calibrate a Vis WS. The trap detectors could also calibrate the UV WS or Ge WS over the part of their spectral ranges that overlap. The Vis WS could do this calibration of the UV WS or Ge WS. The pyroelectric detector is used to extend the spectral power scale beyond the trap detectors calibrated spectral range. The calibrated UV WS and Ge WS are used to calibrate the Vis WS in the spectral region beyond the trap detectors. These transfer steps are described below.

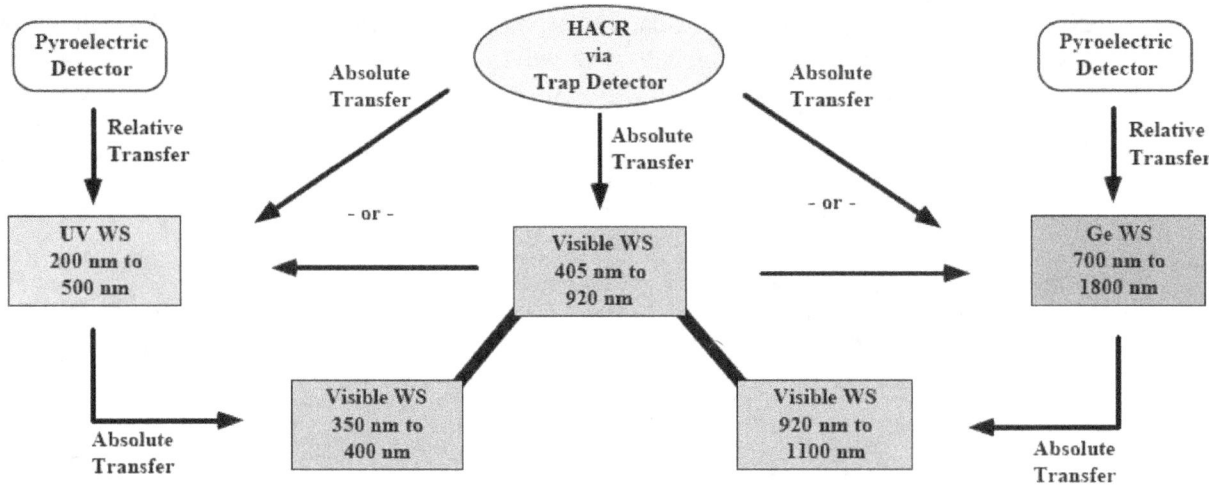

Figure 5.4. NIST spectral power scale propagation chain.

5.2 Traps to Visible Silicon Working Standards

Trap detectors are not used for routine spectral power (responsivity) measurements in the SCFs because the positioning of the trap detectors is time consuming and requires critical alignment to ensure the beam from the Vis/NIR SCF $f/9$ optical system does not overfill any of the trap detector's photodiodes. Also trap detectors have not been available for the UV and near-IR spectral regions, requiring that the spectral power scale be transferred to other detectors. The Vis WS (Hamamatsu S1337-1010BQ) are used for routine measurements since they are easier than the trap detectors to align in the focal plane. Trap detectors can be used to provide higher accuracy measurements when required (over the modeled EQE spectral region).

The four Vis WS are measured against two trap detectors, using the typical measurement setup and routine described later. An automated computer routine is used to determine the center of the active area for each detector. The measurements are then taken at that position. The Vis WS and traps are operated unbiased (the photovoltaic or short-circuit mode) and the signals are measured with calibrated transimpedance amplifiers and DVMs. The amplifier gain for both the working standards and trap detectors is 10^6.

The spectral responsivity is found using eq (3.27) where $S_S = S_T$, the modeled EQE from the HACR. The spectral responsivity of each Vis WS is the weighted mean of the measurement with both trap detectors [40] (see fig. 5.5a.). Figure 4.2 shows that the optical power used for these measurements is typically less than 1 µW.

5.3 Extension to Ultraviolet (200 nm) and Near-Infrared (1800 nm)

Expanding responsivity measurements beyond the modeled EQE spectral region of the trap detectors requires different calibrated detectors. Detectors are required that cover the spectral regions of measurement interest and also overlap with the trap modeled EQE spectral region. If the relative responsivity of the detector were known, comparisons with the trap detectors would then transfer the spectral responsivity scale to the detector. This detector could then be used as a working standard.

One method to extend the spectral range of the responsivity measurements is to use a spectrally flat detector. The spectrally flat detector could be calibrated by comparison with the trap detectors and used as a working standard. A second method is to use a spectrally flat detector to measure the relative responsivity of another detector. This detector could then be calibrated by comparison with the trap detectors and used as a working standard. The latter method was used to extend the spectral range of the responsivity measurements. The transfer method described here uses one broadband, spectrally flat, windowed, pyroelectric detector. The "flatness" of the responsivity is determined by measuring the reflectance from the pyroelectric's surface (absorber) and the window transmission.

The method chosen not only depends on the transfer uncertainty, but also upon the operation of the detectors. Depending on the situation, some detectors, especially spectrally flat detectors, have drawbacks to their use. Cryogenically cooled detectors are often large and heavy, making automated movement of the detectors difficult. And, there is the added complexity of using cryogens. Pyroelectric detectors require a chopped beam (thus additional equipment) and some models have a poor signal-to-noise ratio (SNR) at the power levels typically found with monochromator-based systems. These characteristics may make a detector cumbersome and impractical to use routinely as a working standard.

5.3.1 Pyroelectric Detector

A pyroelectric detector only responds to a change in radiant flux. The operation of pyroelectric detectors thus requires modulated or chopped (ac) optical radiation and lock-in detection. The pyroelectric material is coated with an optically black (flat) material, in this case a gold black. Details of the characteristics and operation of pyroelectric detectors can be found in the literature [41].

The operation of a pyroelectric detector depends upon the spectral absorptance of the gold black material. Using the law of conservation of energy the absorptance $\alpha_p(\lambda)$ is given as

$$\alpha_p(\lambda) = 1 - \rho_p(\lambda) \text{ [unitless]}, \quad (5.1)$$

where $\rho_p(\lambda)$ is the reflectance and the transmittance $\tau_p(\lambda)$ is assumed to be equal to zero. The predicted spectral responsivity $S_p(\lambda)$ of the pyroelectric detector is

$$S_p(\lambda) = \tau_w(\lambda) \cdot \alpha_p(\lambda) \cdot CF_p = \tau_w(\lambda) \cdot (1 - \rho_p(\lambda)) \cdot CF_p \text{ [U·W}^{-1}\text{]}, \quad (5.2)$$

where $\tau_w(\lambda)$ is the pyroelectric detector window transmittance and CF_p is a calibration factor which scales the output signal (in U units) to the optical power received by the pyroelectric detector. The calibration factor could be determined by measurements with the HACR or comparison to the trap detectors. If the calibration factor is set to an arbitrary value, then the relative responsivity $s_p(\lambda)$ is determined. Note that the lower case "s" is used to designate relative responsivity [2].

The reflectance of the pyroelectric detector $\rho_p(\lambda)$ is measured with a modified Perkin-Elmer Lambda-19 spectrophotometer [42]. The window transmittance $\tau_w(\lambda)$ is measured using the NIST transmittance measurement service facility [43]. The window transmittance can also be found by determining the ratio of the signals from two measurements using the SCF, one with and the second without the window in front of a detector.

5.3.2 Responsivity Measurements

From eq (3.27), it can be shown that the relative responsivity $s_x(\lambda)$ of a test detector x can be written

$$s_x(\lambda) = \frac{R_x(\lambda)}{R_p(\lambda)} \cdot s_p(\lambda) \text{ [arbitrary U·W}^{-1}\text{]}, \quad (5.3)$$

where $s_p(\lambda)$ is the relative responsivity of the pyroelectric detector. Equation (5.3) can now be used to determine the relative responsivity of a test detector using the pyroelectric detector as a standard by the same substitution method (measurement equation) described earlier. It should be noted that the pyroelectric detector described here had a poor SNR adding significantly to the measurement uncertainties.

The test detector is then calibrated using the trap or Vis WS detectors over the part of their spectral ranges that overlap. This data is used to scale the relative responsivity of the test detector. Only one point is needed in principal; but, in practice, about 50 different spectral points are used to obtain an average scaling factor to apply to the entire relative responsivity curve.

5.4 Germanium (NIR) Working Standards

The Ge WS relative spectral responsivities are determined using the Vis/NIR SCF and the pyroelectric detector. The weighted average of several measurements are taken, due to the SNR of the pyroelectric detector. The Ge WS responsivity is then measured using the trap detectors over the spectral region from 700 nm to 920 nm. The trap comparison data is the average of three scans during one measurement. This data is used to scale the relative responsivity. Two additional Ge WS, identical to the first two, were measured using the first two Ge WS (comparison using the pyroelectric was not done because of time constraints). See fig. 5.5b and c.

5.5 UV Silicon Working Standards

The relative spectral responsivity of the four UV WS is measured using the UV SCF and the pyroelectric detector. Measurements have been taken with and without the pyroelectric's window. The data without the window is noisier and is not used in determining the UV WS relative spectral responsivity. Even with the window, getting a good SNR requires that many measurements be taken and averaged.

The responsivity of all four UV WS is also measured using all four Vis WS in the Vis/NIR SCF over the spectral region of 405 nm to 500 nm. The average of three scans with each Vis WS was used as the responsivity for each UV WS. This data was used to scale the relative responsivity of the UV WS. See fig. 5.5d and e. The trap detectors were not available, so the Vis WS were used for the scale transfer increasing the relative standard uncertainties from 405 nm to 500 nm by 0.02 %.

5.6 Extension of Visible Silicon Working Standards

The spectral responsivities of the four Vis WS is measured in the Vis/NIR SCF from 925 nm to 1100 nm with the first two Ge WS. The average of three scans with each Ge WS is assigned as the responsivity for each Vis WS in the 925 nm to 1100 nm spectral region. The Vis WS are also measured from 350 nm to 405 nm with the UV WS in the Vis/NIR SCF. The average responsivity of three scans with all four UV WS is assigned to each Vis WS. Figure 5.5f shows the Vis WS scale extension measurements. The uncertainties are significantly higher in these regions, compared to the trap spectral region, due to the pyroelectric detector SNR and the additional transfer measurements.

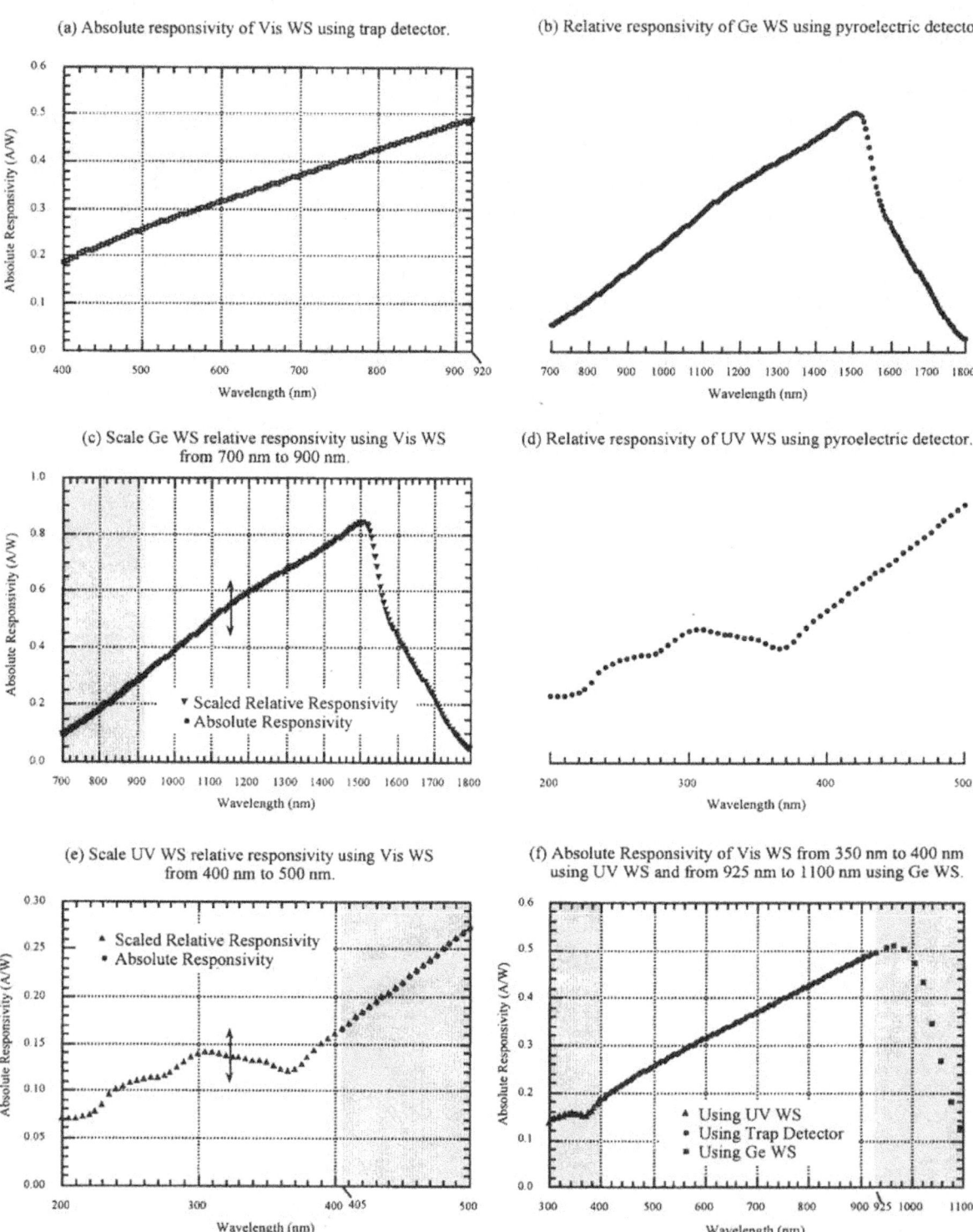

Figure 5.5. NIST detector spectral responsivity scale realization. Shaded areas indicate regions of spectral overlap for calibration transfer.

6. Calibration Procedures and Computer Automation

6.1 Spectral Responsivity

This section describes the spectral responsivity measurement procedures for both the UV SCF and Vis/NIR SCF. These correspond to Service ID numbers 39071S through 39075S. The primary difference between the operation of the two SCFs is that the UV SCF can accommodate only one test detector per responsivity measurement run, and the Vis/NIR SCF is set up for two test detectors. The spatial uniformity of the responsivity for new photodiodes, which is included in 39071S and 39073S, is also measured as described in section 6.2.

6.1.1 Calibration Procedures

New photodiodes are visually inspected for defects and placed in one of the specially designed fixtures described later in section 9.5. Each photodiode has a serial number engraved on the fixture. The photodiode dynamic impedance (shunt resistance) is measured with an HP 4145A I-V plotter to confirm the manufacturer's specification. The shunt resistance of the photodiode needs to be 1000 times greater than the input impedance of the transimpedance amplifier for it not to deviate from linear operation. Typically, an amplifier gain of 10^6 is used, requiring that the photodiode shunt resistance must be greater than 10 kΩ. If needed, the diode window is cleaned with lens tissue and spectral-grade acetone before any optical measurements are made. Ethanol has also been used by others to clean photodiodes and optical windows [44].

The spectral responsivity is measured by direct comparison to two working standards using the "substitution method with monitor" described in section 3.2. The two working standards are selected from a randomly generated weekly schedule. The test detector(s) and working standard detectors are aligned perpendicular to the optical axis by using the He-Ne laser as the monochromator source and retroreflecting the He-Ne beam back onto itself. The appropriate broadband source used for the measurement is chosen and the detector positioned at the focal plane of the SCF exit optics. An automated computer routine centers the active area of each detector on the optical axis.

The typical comparison measurement consists of scanning the monochromator through the desired spectral range at wavelength intervals of 5 nm for each detector. This spectral scanning process is repeated three times. The test detector(s) and Vis WS are operated unbiased (the photovoltaic or short-circuit mode) and the signals are measured with calibrated transimpedance amplifiers and DVMs. Figure 4.2 shows that the optical power used for these measurements is typically less than 1 μW. The test to monitor (eq (3.23)) and working standard to monitor (eq (3.24)) ratio data are stored on the computer for later analysis to determine the spectral responsivity. The standard deviation of the test (or working standard) and monitor detector ratios is less than either individual signal standard deviation. This is the result of simultaneously sampling both the test (or working standard) and monitor detector. Examples of the typical signals from a Hamamatsu S1337 and the monitor photodiode are shown in figure 6.1. Also, shown are the signal ratios and the relative standard deviations of each signal and ratio. Figure 6.1 shows the ratio relative standard deviation is lower than either individual signal standard deviation.

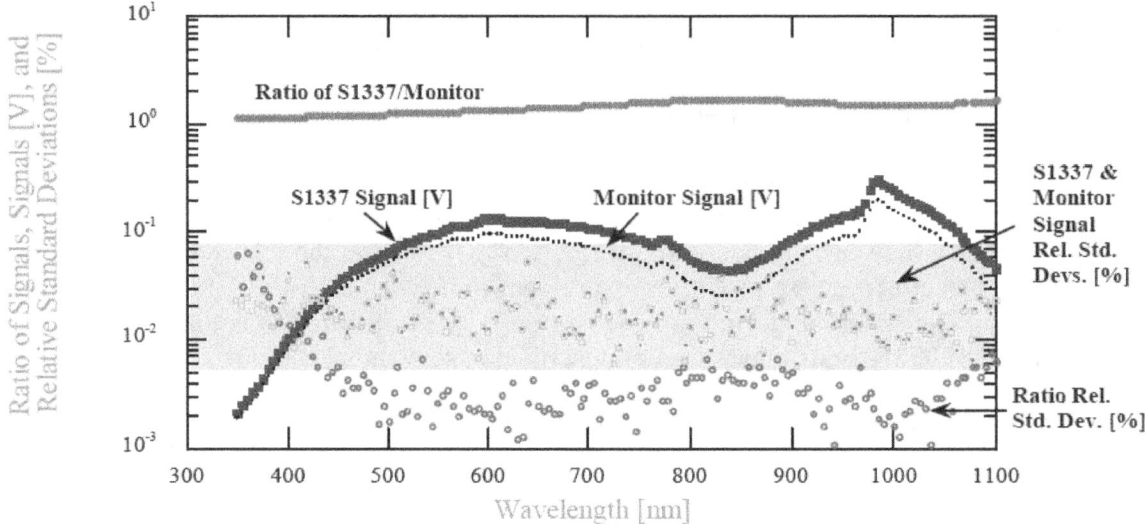

Figure 6.1. Typical signals from Hamamatsu S1337 and monitor photodiodes. The S1337 and monitor signals are shown as closed squares and triangles, respectively. Also shown are the signal ratios as closed circles. All curves are the means of 10 samples. The relative standard deviations of the S1337 and monitor signals are shown respectively as open squares and triangles in the shaded area. The relative standard deviations of the ratios are shown as open circles and for measurements above 400 nm are lower than the individual signal standard deviations. Note: The S1337 and monitor signal standard deviation curves are nearly indistinguishable because the source noise rather than detector noise dominates the measurement.

The laboratory environment (temperature, humidity, etc.) is monitored and recorded at the start of each scan although this data is not used to correct the measurement results. The temperature of specially designed detectors that have temperature sensors built into their housings can also be recorded. The average temperature during the measurements is then reported.

The spectral responsivity is determined by using eq (3.27) for each test and working standard detector spectral scan combination. The reported spectral responsivity of the test detector is the weighted mean [40] of all the scans with both working standard detectors. Examples of typical photodiode spectral responsivities are shown in figure 6.2.

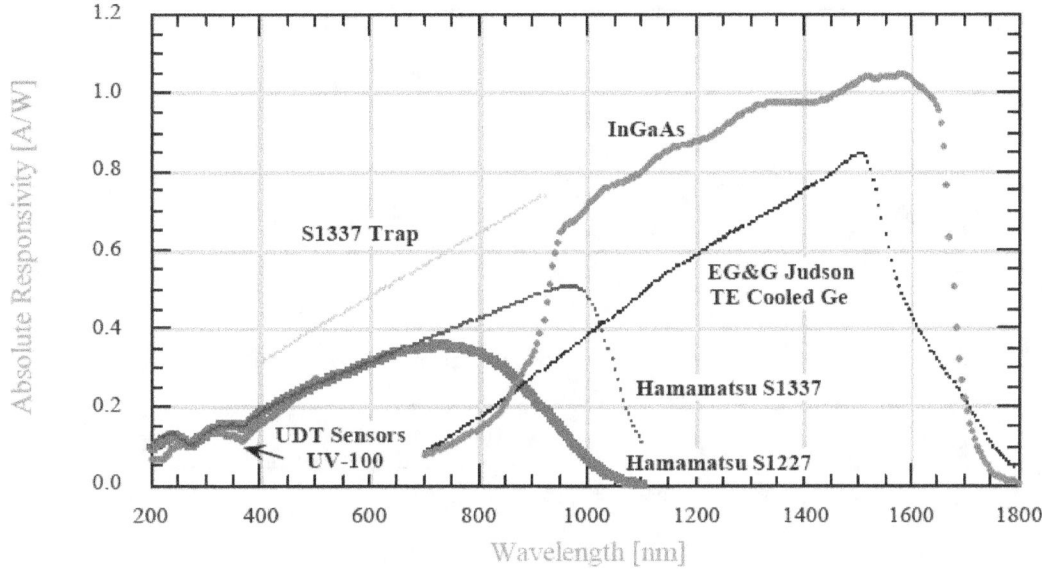

Figure 6.2. Spectral responsivities of typical Si, InGaAs, and Ge photodiodes.

6.1.2 Quantum Efficiency

The quantum efficiency is the photon-to-electron conversion efficiency of a photoelectric detector. The quantum efficiency of a detector is often required for particular applications. There is a simple calculation to convert spectral responsivity [A·W⁻¹] to external quantum efficiency. The external quantum efficiency EQE is given by [10]

$$EQE(\lambda) = \frac{I(\lambda) \cdot h \cdot c}{\Phi(\lambda) \cdot n \cdot e \cdot \lambda} , \quad (6.1)$$

where I is the photocurrent (output current minus the dark output), h is Planck's constant, c is the velocity of light, Φ is the input radiant flux (power), n is the index of refraction of air, e is the elementary electronic charge, and λ is the spectral wavelength. Substituting for the constants h, c, n, and e gives

$$EQE(\lambda) = 1239.85 \cdot \frac{I(\lambda)}{\Phi(\lambda) \cdot \lambda} = 1239.85 \cdot \frac{S(\lambda)}{\lambda} , \quad (6.2)$$

where $S(\lambda) = I(\lambda)/\Phi(\lambda)$ is the spectral responsivity [A·W⁻¹], and for convenience, λ is in nm. Examples of typical photodiode EQEs are shown in figure 6.3.

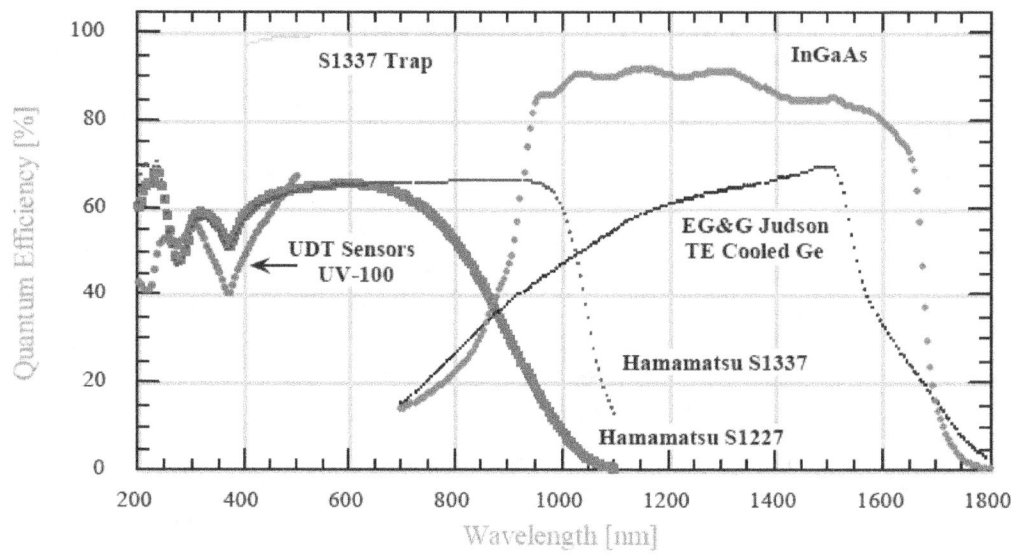

Figure 6.3. Quantum efficiencies of typical Si, InGaAs, and Ge photodiodes.

6.2 Spatial Uniformity

This section describes the procedures used in NIST Service ID numbers 39071S, 39073S, and 39081S to determine the responsivity spatial uniformity of detectors. The spatial uniformity measurement procedures are the same for both the UV SCF and Vis/NIR SCF. The major difference between the two is the beam size. The UV SCF beam diameter is 1.5 mm and the Vis/NIR SCF beam diameter is 1.1 mm. Also, only one detector can be measured at a time in the UV SCF while up to four detectors can be measured in the Vis/NIR SCF.

The spatial uniformity is measured at 350 nm for the UDT Sensors UV100 photodiodes (39071S). The spatial uniformity for the Hamamatsu S1337-1010BQ and, more recently, the Hamamatsu S2281 photodiodes is measured at 500 nm (39073S). UV100 photodiodes are not issued if the responsivity nonuniformity (i.e., slope) is greater than 1 % over the active area or greater than 0.5 % within the center 50 % of the active area or if a discontinuity (a peak or valley) in the responsivity greater than 0.5 % is found within the center 90 % of the active area. Hamamatsu S1337-1010BQ and S2281 photodiodes are not issued if the nonuniformity is greater than 0.5 % over the active area or greater than 0.25 % within the center 50 % of the active area or if a discontinuity greater than 0.25 % is found within the center 90 % of the active area.

In the early 1990's, it was found that many silicon photodiodes have a significant change in the uniformity as a function of wavelength, particularly as the wavelength approached the bandgap (1100 nm). Nonuniformity is due to inhomogeneity in the photodiode material - typically inhomogeneity in surface recombination centers at shorter wavelengths [45, 46] and bulk recombination centers at longer wavelengths [47]. The nonuniformity near the bandgap has also been related to the (non)uniformity of the bonding material's reflectance [48]. Because the semiconductor is almost transparent near the bandgap, changes in the spatial reflectivity of the bonding material affect the amount of light reflected and therefore the responsivity of the photodiode.

Since 1993, the uniformity of the S1337-1010BQ and the S2281 diodes are also measured at 1000 nm [49]. Photodiodes are not issued if the nonuniformity is greater than 1 % over the active area or greater than 0.5 % within the center 50 % of the active area. Also, if a discontinuity greater than 0.5 % is found within the center 90 % of the active area.

S1337-1010BQ photodiodes with a significant change in uniformity between 500 nm and 1000 nm are seen less frequently in recently manufactured diodes. Diodes issued by NIST prior to the discovery of this effect in 1993 were not measured at 1000 nm. When these diodes are resubmitted for measurement (39074S), the uniformity is checked at 1000 nm; and the customer is notified if a significant nonuniformity is found.

6.2.1 Measurement Method and Calibration Procedure

The detectors are aligned as described above in section 6.1. The typical measurement consists of setting the monochromator to the desired wavelength and, for a 1 cm^2 test detector, scanning a 12 mm x 12 mm area in 0.5 mm steps. The test detector is operated unbiased (the photovoltaic or short-circuit mode); and the signal is measured with a calibrated transimpedance amplifier and a DVM. The typical amplifier gain for the test detector is 10^6. Figure 4.2 shows the optical power used for these measurements is typically less than 1 µW. The test to monitor (eq (3.23)) ratio data is stored on the computer for later analysis.

The scans are always in the same horizontal direction; and the vertical direction is reversed in a "raster" scan, starting in the upper left corner moving to the lower right corner of the photodiode. The scan is large enough for the beam to move completely off of the active area (or onto an aperture). Scanning horizontally in only one direction puts the stage drive against the same side of the drive screw. For the vertical, gravity keeps the stage always against the same side of the screw. This reduces hysteresis in the movement of the stages.

The laboratory environment (temperature, humidity, etc.) is monitored and recorded at the beginning and end of each scan, although this data is not used to correct the measurement results. The temperature of specially designed detectors that have temperature sensors built into their housings can also be recorded. The average temperature during the measurements is then reported.

The test to monitor detector ratios are normalized to the mean of the center ratios. The reported spatial uniformity figure is constructed from these normalized ratios. Figure 6.4 shows the spatial uniformities of the central portion of typical Hamamatsu S1337-1010BQ and S2281 photodiodes at 500 nm and 1000 nm. Similar spatial uniformity scans are shown in figure 6.5a for UV100 and figure 6.5b, c, and d for Judson EG&G thermoelectrically cooled Ge photodiodes.

6.2.2 Limitations

The size of the active area to be measured has to be significantly larger than the beam size since the beam is "clipped" (or vignetted) at the edges of the active area or aperture in front of the detector. Also the spatial shape of the optical beam is assumed constant during the measurement.

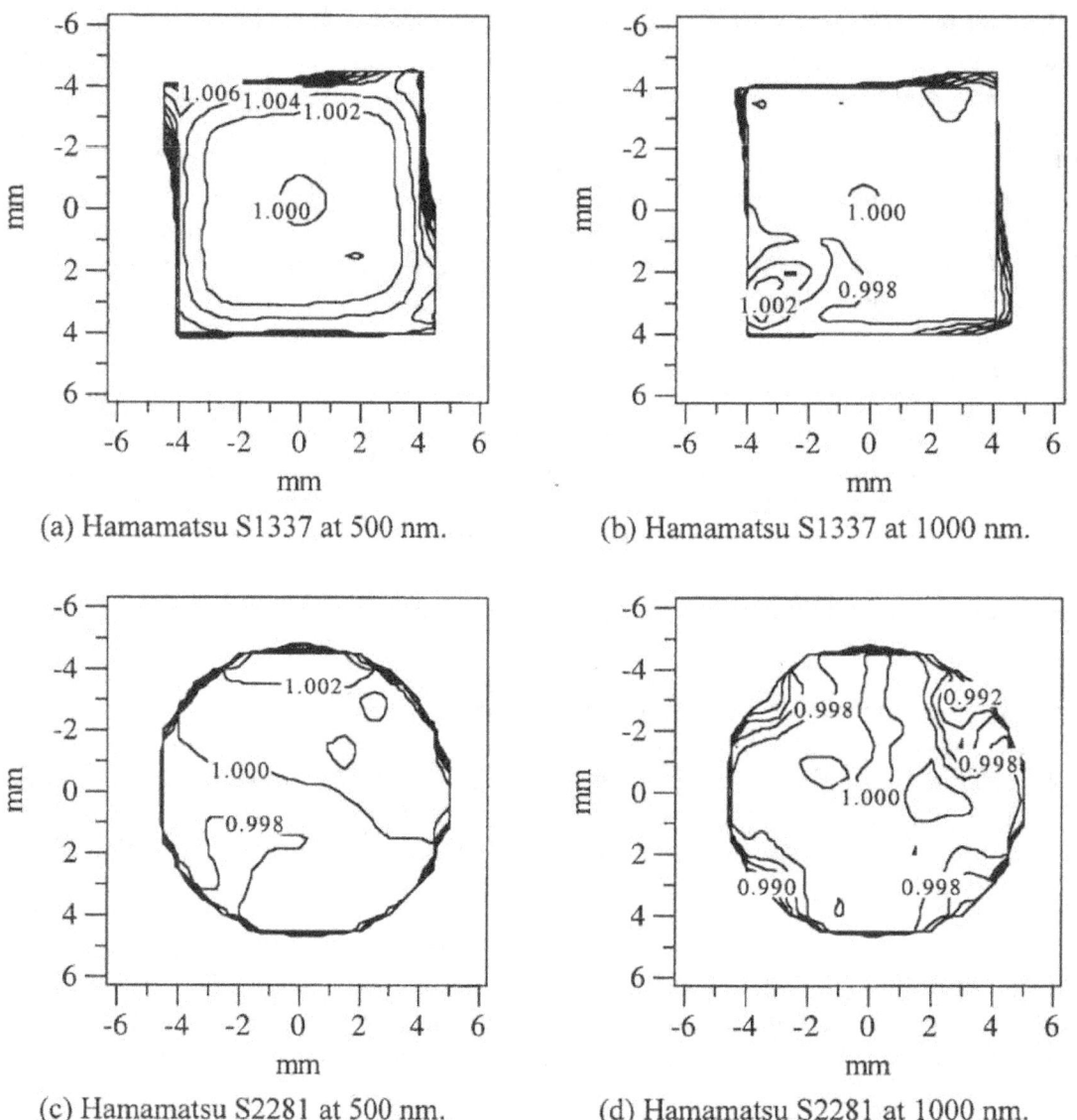

(a) Hamamatsu S1337 at 500 nm.

(b) Hamamatsu S1337 at 1000 nm.

(c) Hamamatsu S2281 at 500 nm.

(d) Hamamatsu S2281 at 1000 nm.

Figure 6.4. Spatial uniformities of typical Hamamatsu S1337 and S2281 photodiodes. The responsivities are normalized to the center values.

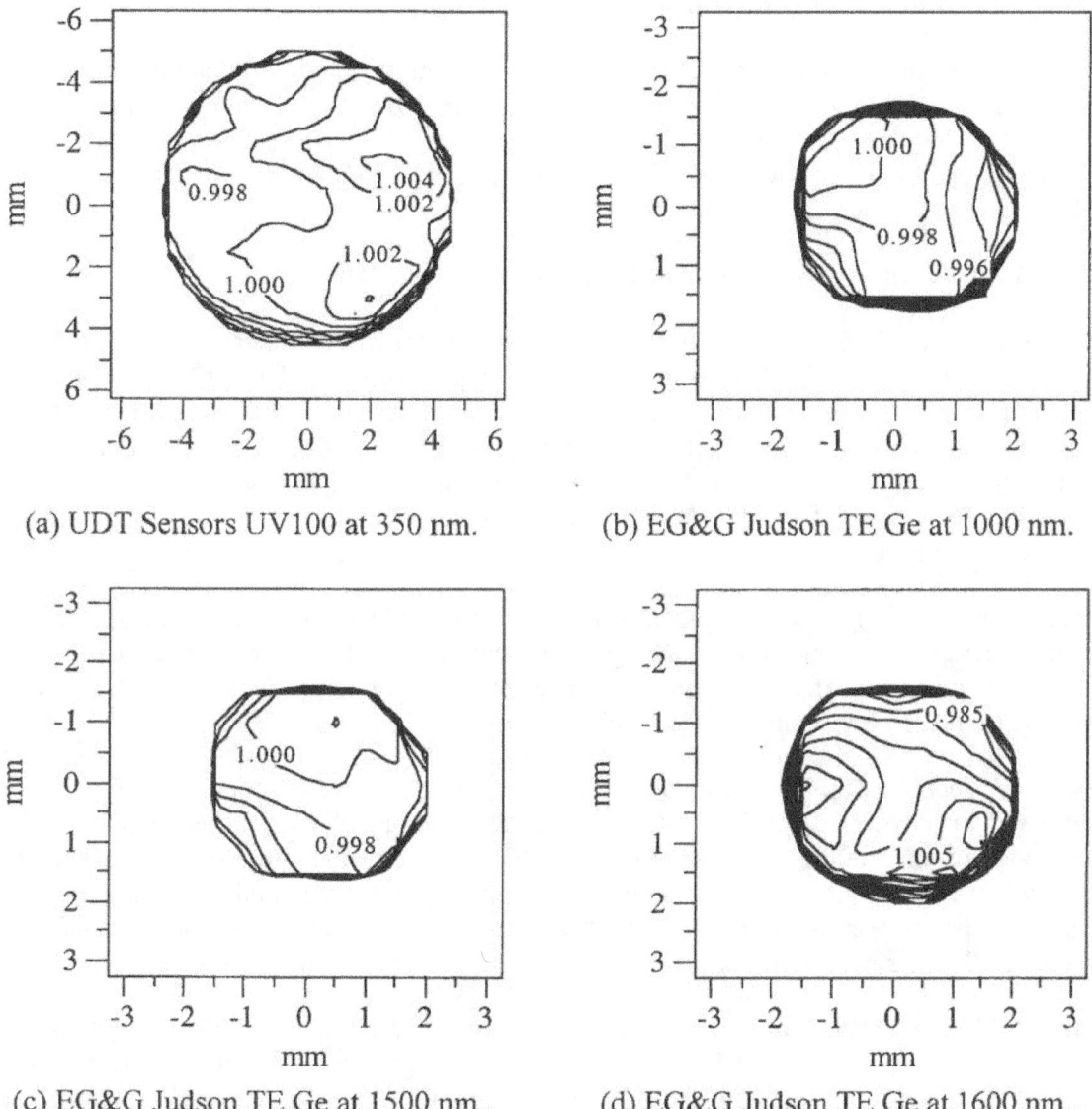

(a) UDT Sensors UV100 at 350 nm.

(b) EG&G Judson TE Ge at 1000 nm.

(c) EG&G Judson TE Ge at 1500 nm.

(d) EG&G Judson TE Ge at 1600 nm.

Figure 6.5. Spatial uniformities of typical UDT Sensors UV100 and EG&G Judson Ge photodiodes. The responsivities are normalized to the center values.

6.3 Computer Automation

This section describes the computer automation of the UV and Vis/NIR comparator facilities. The automated equipment and computer software used for typical measurements are briefly described. Automating the equipment (which sometimes requires modification) and writing, testing, and documenting the software is a major undertaking. The operation of the facilities would not be practical without this high degree of computer automation.

6.3.1 Computer Automated Equipment

Block diagrams that include current and future computer controlled equipment for the Vis/NIR SCF and UV SCF are shown in figures 6.6 and 6.7, respectively. All of the equipment is controlled by one computer via an IEEE-488 (GPIB) bus. This computer stores all of the data and the analyzed responsivity files. A Local Area Network (LAN) connects this computer with several others which are used for further data comparisons and writing test reports; allowing the SCF control computer to be dedicated to taking measurements. Weekly backups of the data files are sent to a second computer.

Several key components of the SCFs are controlled by commercial servo motor controllers: the horizontal and vertical (x,y) translation stages, source section, and the rotary stage in the UV SCF. The wavelength drive of the Cary-14 monochromator used with the Vis/NIR SCF was modified and is also driven by a computer controlled servo motor.

All of the DVMs, lock-in amplifiers, and multiplexers are computer controlled. A digital I/O module addressed over the IEEE-488 bus signals the shutter controllers to open or close. A commercial laboratory environmental monitor records the laboratory and enclosure temperatures, the humidity and barometric pressure, and the electrical power line voltage and frequency. Some detectors have temperature monitoring circuitry that produces a voltage signal proportional to their temperature. These signals can also be multiplexed via computer control to the DVM.

In the future the TE temperature controllers and transimpedance amplifiers will also be controlled via IEEE-488 bus and computer. The Vis/NIR SCF wavelength encoder and display will be upgraded, allowing communication with the controlling computer. A new monochromator will soon be incorporated in UV SCF which will have computer selectable order sorting filters.

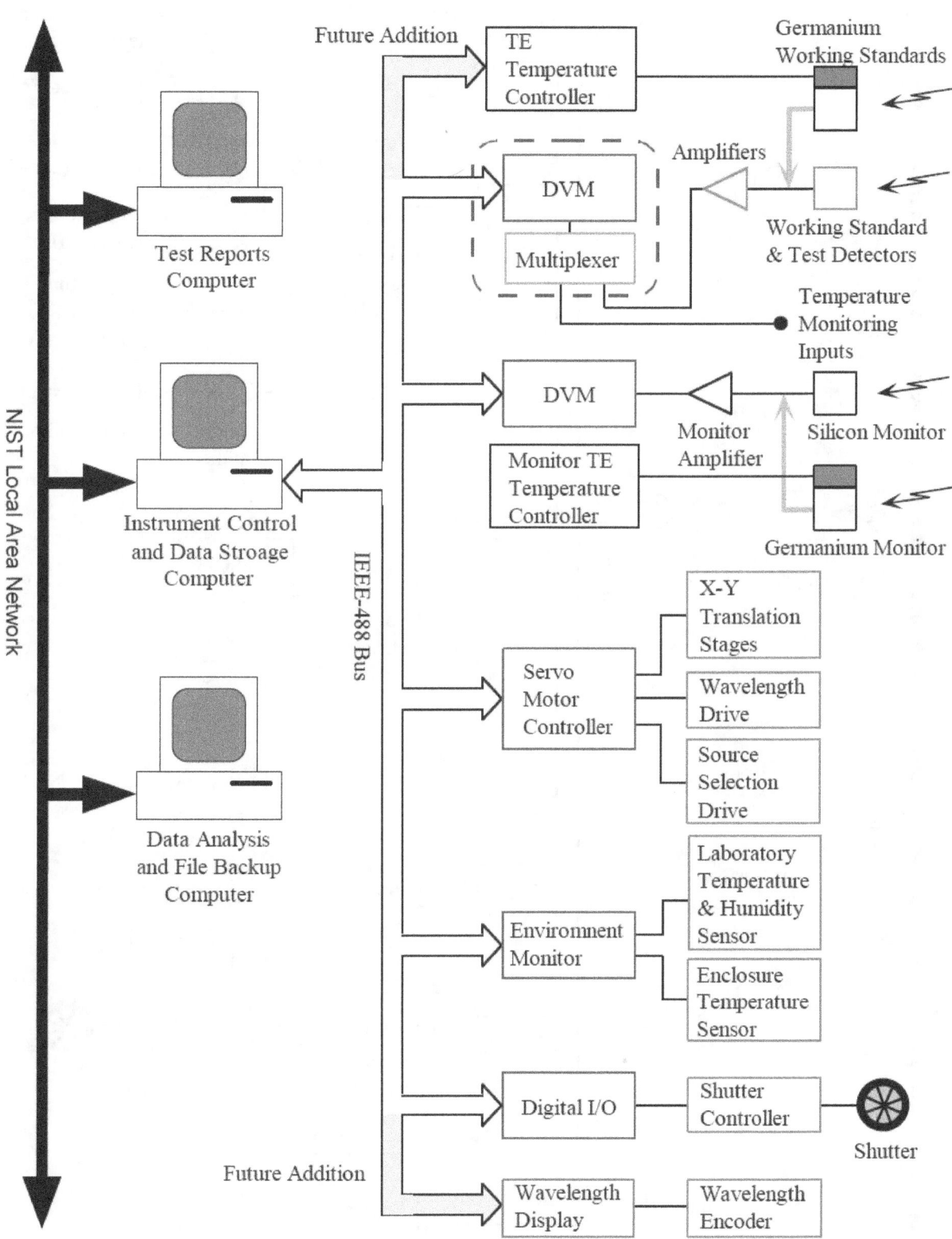

Figure 6.6. Vis/NIR SCF computer control block diagram.

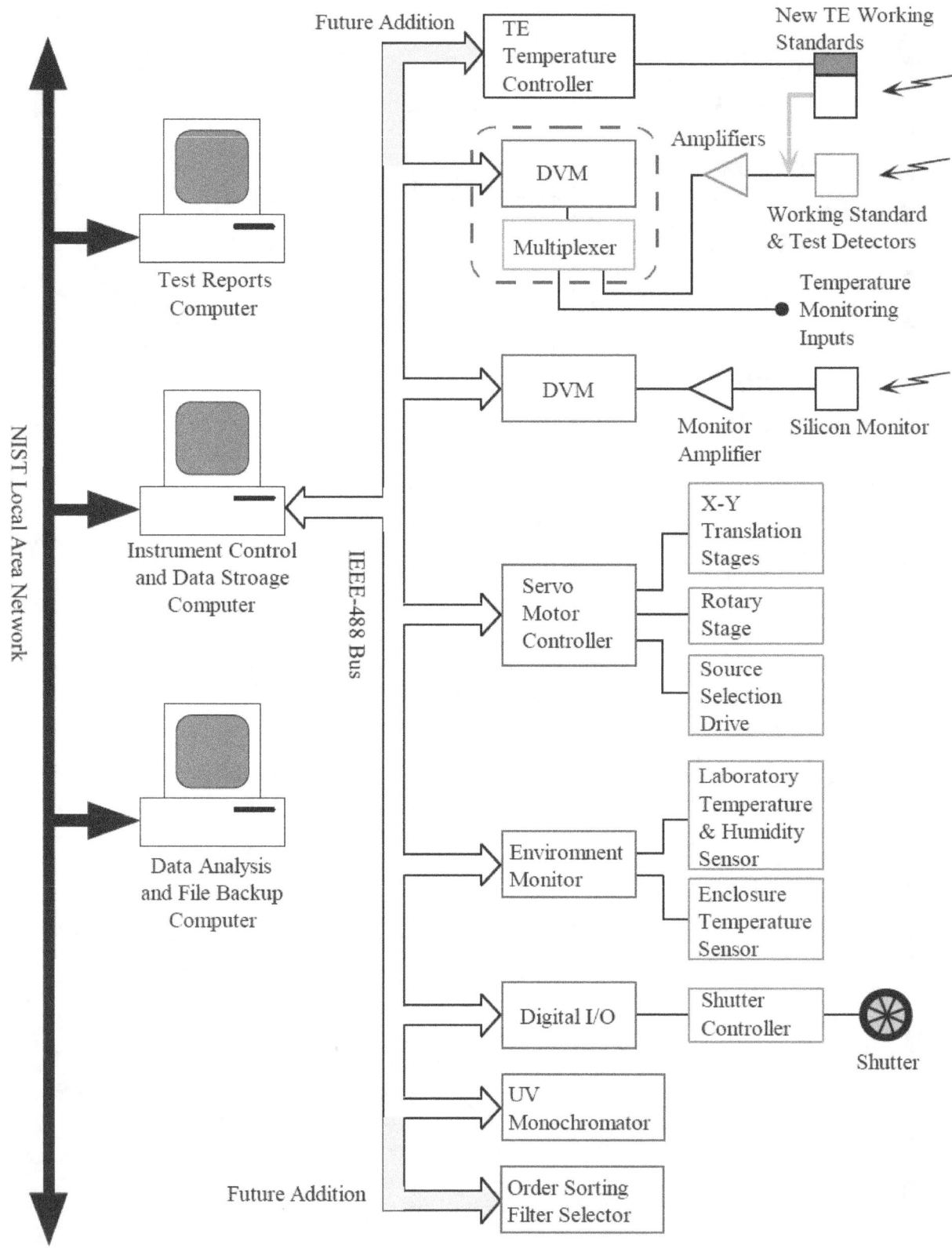

Figure 6.7. UV SCF computer control block diagram.

6.3.2 Computer Calibration Programs

SCF Setup Program

This program performs three functions: initialize the SCF instruments, align the test and working standard detectors, and verify the signal levels from the detectors before starting measurements. The operator uses this program interactively (via the front panel controls and translation stage joystick) to align a new detector placed in the SCF. This program updates a file that contains the detector names and translation stage (x,y) positions for use later by other computer programs.

Spectral Scanning Program

This program takes data from one to four detectors (test and working standard detectors), stores the data on hard disk, and prints the results. The analysis program can be set to automatically run when the measurements finish. The operator sets several detector parameters such as the time constant and amplifier gain, whether to use a detector (position), and whether the detector is a working standard. Other operator inputs are the spectral range start and stop wavelengths, spectral step size, number of samples at each wavelength, number of (repeat) scans for the measurement, and which SCF source to use. The operator can also input specific comments about the measurement.

After the parameters have been entered, the program positions the first detector into the beam and starts taking data (see fig. 6.8). The program determines the test (measurement) number and creates unique filenames for each detector using this number and the detector's name entered in the SCF setup program. The program updates the computer screen with the current scan number, the detector name and position for the detector currently being measured, filename, and the pathname to the data stored on the hard disk. The program also graphs the test detector to monitor ratio (eq (3.23)), the working standard detector to monitor ratio (eq (3.24)), and the percent standard deviation of the ratios. The program saves the "raw" ratio data files on the hard disk after each detector spectral scan. After completion of the measurement, a record is printed of the operator inputs, graph of the detector to monitor ratios, and graph of the percent standard deviation of the ratios.

The "raw" ratio data files are two dimensional ASCII text arrays that are tab delimited and consist of "header" lines and data lines. The "header" lines contain information about the measurement parameters, laboratory environment (e.g., temperature), and detector temperature (if provided) or auxiliary thermistor reading. The data columns are: (1) wavelength, (2) mean ratio of the (test or working standard) detector signal to the monitor signal, (3) the percent standard deviation of those ratios, (4) mean detector signal, (5) detector percent standard deviation, (6) mean monitor signal, and (7) monitor signal percent standard deviation. (See fig. 6.1.) If more than one scan is taken then the environmental and detector temperature information and additional data are added at the end of the data file after each scan. If selected by the operator, the spectral responsivity calculation program is automatically executed after all the spectral scans have completed.

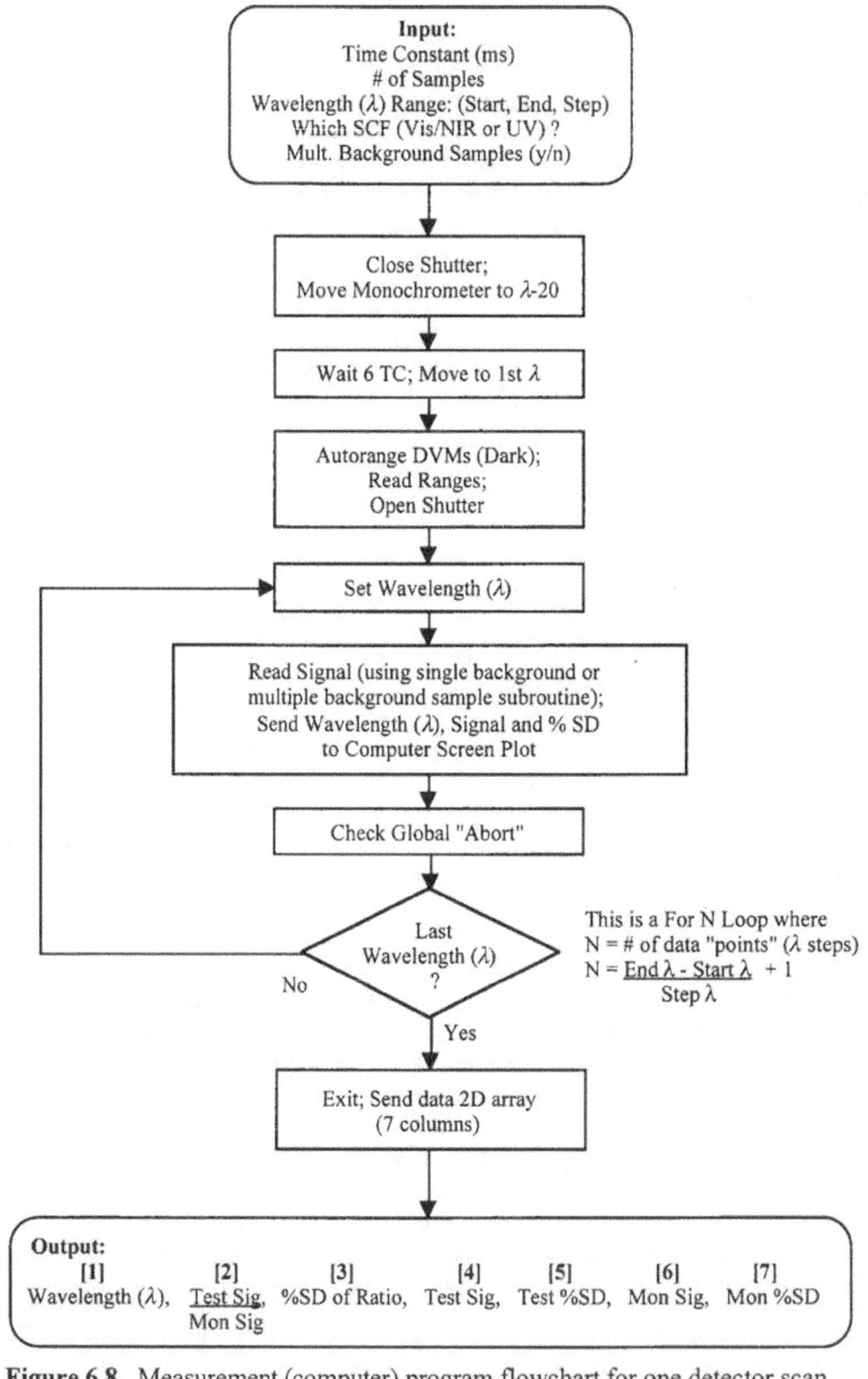

Figure 6.8. Measurement (computer) program flowchart for one detector scan.

Spectral Responsivity Calculation Program

This program calculates the spectral responsivity for the test detector(s), stores the results on the hard disk, and prints the (weighted) mean responsivity and relative measurement uncertainty (graphs), detector filenames, and control settings. The inputs to the program are the test and working standard detector "raw" ratio data filenames, which working standard reference (calibration) data to use, and whether a NIST transimpedance amplifier or other calibrated amplifier was used with the test detector.

The gain range and detector position in the SCF are extracted from the data files for the test and working standard detectors. This is used to select the corrected (calibrated) gain factor for the appropriate NIST amplifier. For detectors with internal amplifiers the gain range is multiplied by the customer supplied amplifier gain correction factor to determine the corrected gain factor. The program next reads and formats the ratio data files and the working standard reference data file. The program determines the responsivity for a test detector by using eq (3.27) for each test and working standard detector spectral scan combination. Then if more than one spectral scan was taken a weighted average of the mean is calculated [40] using the standard deviations from each scan as the weighting factor. The weighted average uncertainty is also calculated.

If the responsivity data is negative (because of reversed polarities on the photodiodes) the responsivity data is inverted, i.e., $-S(\lambda)$. Then the spectral responsivity and relative uncertainty are graphed on the computer monitor. The detector temperature monitor (or auxiliary thermistor) data for all the scans are averaged for each test detector. The output filename is unique and constructed very similar to the "raw" data filename using the test number and test detector name. The responsivity output file is created as a tab delimited ASCII text array. The file consists of "header" lines and data lines. The "header" lines contain information about the test number, detector (name), test date, working standard(s) and reference data, internal amplifier gain and correction factor (if applicable), average detector temperature, and column headings. The data columns are: (1) wavelength, (2) weighted average spectral responsivity, (3) relative uncertainty. After the measurement is complete a record of the operator inputs and graphs of the weighted average spectral responsivity and relative uncertainty are printed.

Spatial Scanning Program

This program spatially scans the SCF beam across the active area of the detector by using the x,y translation stages to translate the detector. The data is stored on hard disk, and the measurement results are printed when the program finishes execution. One detector in the UV SCF and up to four detectors in the Vis/NIR SCF can be measured at one time. The measurements can be programmed to repeat at different wavelengths.

The program inputs are the scan wavelength, number of samples at each data point, detector parameters such as the time constant and amplifier gain, whether to use a detector (position), and operator comments. Also input are the spatial scan horizontal (x) and vertical (y) dimensions and step sizes. The program determines the test number and creates a unique test filename with the test number, detector name, and wavelength.

The program moves the detector to the first x and y position and takes the detector to monitor signal ratio measurement (eq (3.23)). The program scans the detector from top to bottom and left to right. That is, the x,y movement is from the top left to the bottom right (or top right depending on the number of columns measured). The program displays a *quasi* three dimensional graph of the signal ratios. The absolute value of the ratio data is plotted in case the polarity of the diode is reversed. This graph is updated after each data point measurement. Also displayed are the current horizontal and vertical positions. The ratio and percent standard deviation data from each x,y position are combined into arrays and added to the output file saved on the hard disk. After the spatial scan is complete a record of the operator inputs and a spatial graph of the signal ratios is printed.

The responsivity spatial uniformity output data file is an ASCII text array that is tab delimited. The file consists of "header" lines and data lines. The "header" lines contain information about the measurement parameters, operator comments, laboratory temperature, humidity, etc., and detector temperature (if provided). The data lines consist of y rows and x columns of the test detector to monitor detector signal ratios. The first data point (array cell) corresponds to the upper left x,y scan position. That is, the array corresponds to looking at the detector where the left side of the array is the left side of the detector, the top of the array is the top of the detector, etc. Following the ratio data array is a similar array containing the corresponding relative standard deviations.

7. Uncertainty Assessment

The assessment of the uncertainties for the detector responsivity measurement is explained in this section. First the uncertainty is evaluated for the measurement equation developed earlier in this document. In addition to the uncertainty terms that come directly from the measurement equation there are indirect terms due to the wavelength uncertainty of the monochromator, long-term stability of the working standards, and the assumptions and approximations made during the development of the measurement equation. Second, the uncertainty for each group of working standards and the transfer to customer detectors is given in detail. The uncertainty analysis follows the method outlined in Ref. [7]. A general discussion of the sources of error in radiometry can be found in Refs. [50 and 51]. A detailed explanation of the evaluation and expression of measurement uncertainty is given in Ref. [52].

7.1 Uncertainty Components

In general, a measurement result y can be expressed as a functional relationship f of N input quantities x_i given by,

$$y = f(x_1, x_2, ..., x_N). \tag{7.1}$$

The combined standard uncertainty $u_c(y)$ is given by the law of propagation of uncertainty as the following sum,

$$u_c(y) = \sqrt{\sum_{i=1}^{N}\left(\frac{\partial f}{\partial x_i} \cdot u(x_i)\right)^2}, \qquad (7.2)$$

where the partial derivatives $\partial f / \partial x_i$ are the sensitivity coefficients and $u(x_i)$ are the standard uncertainties of each input x_i which are assumed to be uncorrelated. This method is also called the root-sum-of-squares or "RSS" method.

When the functional relationship involves products and quotients, as the measurement equation described in this document, the relative combined standard uncertainty $u_c(y)/y$ is more convenient, and is given by

$$u_{c,r}(y) \equiv \frac{u_c(y)}{y} = \sqrt{\sum_{i=1}^{N}\left(\frac{1}{y} \cdot \frac{\partial f}{\partial x_i} \cdot u(x_i)\right)^2}, \qquad (7.3)$$

where $(1/y) \cdot (\partial f / \partial x_i)$ is the relative sensitivity coefficient. The expanded uncertainty U is obtained by multiplying $u_c(y)$ by a coverage factor k,

$$U = k \cdot u_c(y), \qquad (7.4)$$

where k is chosen on the basis of the level of confidence desired. Replacing $u_c(y)$ with the relative combined standard uncertainty $u_{c,r}(y)$ gives the relative expanded uncertainty $U_r \equiv U/y$. The coverage factor $k = 2$ was used in this document and it is assumed that the possible estimated values of spectral responsivity are approximately normally distributed with approximate standard deviation u_c, thus, the interval defined by U has a level of confidence of approximately 95 %.

To help clarify the following discussion of uncertainties, an arbitrary uncertainty $u_0(S_{ws})$ in the working standard spectral responsivity is defined. The uncertainty $u_0(S_{ws})$ can be derived directly from the measurement equation in eq (3.27), repeated below, by using the propagation of standard uncertainty relationship in eq (7.2). The measurement equation is

$$S_{ws} = \frac{R_{ws}}{R_s} \cdot \frac{G_s}{G_{ws}} \cdot S_s \ [\text{A} \cdot \text{W}^{-1}]. \qquad (7.5)$$

The uncertainty $u_0(S_{ws})$ is

$$u_0(S_{ws}) = \left[\left(\frac{\partial S_{ws}}{\partial R_{ws}} \cdot u(R_{ws})\right)^2 + \left(\frac{\partial S_{ws}}{\partial R_s} \cdot u(R_s)\right)^2 + \left(\frac{\partial S_{ws}}{\partial G_s} \cdot u(G_s)\right)^2 \right.$$
$$\left. + \left(\frac{\partial S_{ws}}{\partial G_{ws}} \cdot u(G_{ws})\right)^2 + \left(\frac{\partial S_{ws}}{\partial S_s} \cdot u(S_s)\right)^2\right]^{1/2} \ [\text{A} \cdot \text{W}^{-1}]. \qquad (7.6)$$

The relative uncertainty $u_0(S_{ws})/S_{ws}$ is

$$\frac{u_0(S_{ws})}{S_{ws}} = \sqrt{\left(\frac{u(R_{ws})}{R_{ws}}\right)^2 + \left(\frac{u(R_s)}{R_s}\right)^2 + \left(\frac{u(G_s)}{G_s}\right)^2 + \left(\frac{u(G_{ws})}{G_{ws}}\right)^2 + \left(\frac{u(S_s)}{S_s}\right)^2}, \qquad (7.7)$$

where $u(R_{ws})$ is the standard deviation of the mean of i ratios of the working standard and the monitor, $u(R_s)$ is the standard deviation of the mean of i ratios of the calibration standard and the monitor, $u(G_s)$ is the calibration standard amplifier gain uncertainty, $u(G_x)$ is the working standard amplifier gain uncertainty, $u(S_s)$ is the calibration standard uncertainty, and i is the number of ratio measurements taken.

The calibration standard uncertainty, $u(S_s)$, is determined previously by the HACR transfer measurements. The amplifier gains, G_s and G_x, are calibrated using a precision current source and DVM. The amplifier gain uncertainties, $u(G_s)$ and $u(G_x)$, are the RSS of the uncertainties from the precision current source and DVM.

The first two terms contribute uncertainty due to the measurement statistics (Type A uncertainties). The remaining terms are Type B uncertainties. If more than one scan is taken during a measurement (typically three scans are taken), then a weighted average of the mean, R_w, of the ratio $R = R_{ws}/R_s$ is calculated [40] using the RSS of $u(R_{ws})$ and $u(R_s)$ from each scan as weighting factors. An uncertainty term $u(R_w)$ is calculated for the weighted average and is an indication of the measurement repeatability.

There is also a statistical uncertainty term $u(R_r)$ reflecting the reproducibility of the measurements (e.g., setting up the measurement again the next day). The relative uncertainty term, $u(R_r)/R_r$ is combined by RSS with $u(R_w)/R_w$ to get the relative measurement uncertainty, $u(R_m)/R_m$. The relative measurement uncertainty contains the temperature and uniformity variations since it is the result of repeated measurements at different temperatures and alignments and is an indication of the measurement reproducibility.

By calculating the ratios of the test to monitor and standard to monitor signals, we take into consideration the correlation of the signals due to source fluctuations, thus the covariance is not included in the R uncertainty term. The standard deviation of the ratio of the detector to monitor signals is less than the standard deviation of either by themselves, as shown in figure 6.1.

The combined uncertainty $u_c(S_{ws})$ in the spectral responsivity of the working standard is determined by calculating the RSS of the uncertainty terms thus far discussed and the uncertainty terms from the assumptions and approximations made in the measurement equation and so-called "indirect" effects that are not explicitly shown in the measurement equation as expressed in eq (7.5) such as, the monochromator wavelength uncertainty, DVM uncertainty, and long-term stability of the working standards.

The relative combined uncertainty $u_c(S_{ws})/S_{ws}$ is

$$\frac{u_c(S_{ws})}{S_{ws}} = \left[\left(\frac{u(R_m)}{R_m}\right)^2 + \left(\frac{u(G_s)}{G_s}\right)^2 + \left(\frac{u(G_{ws})}{G_{ws}}\right)^2 + \left(\frac{u(S_s)}{S_s}\right)^2 + \left(\frac{u(S_\lambda)}{S_\lambda}\right)^2 \right.$$
$$\left. + \left(\frac{u(V)}{V}\right)^2 + \left(\frac{u(S_{lt})}{S_{lt}}\right)^2 + \left(\frac{u(\gamma)}{\gamma}\right)^2 + \left(\frac{u(R_{sl})}{R_{sl}}\right)^2 + \left(\frac{u(S_{bw})}{S_{bw}}\right)^2\right]^{1/2}, \quad (7.8)$$

where $u(S_\lambda)/S_\lambda$ is the monochromator wavelength relative uncertainty, $u(V)/V$ is the DVM relative uncertainty, $u(S_{lt})/S_{lt}$ is the relative uncertainty due to the long-term stability of the working standards, $u(\gamma)/\gamma$ is the scaling factor relative uncertainty for the relative responsivity measurements, $u(R_{sl})/R_{sl}$ is the relative uncertainty due to stray light, and $u(S_{bw})/S_{bw}$ is the relative uncertainty due to the bandwidth-effect. These uncertainty terms are explained below.

7.1.1 "Indirect" Uncertainty Components

Several uncertainty components are not explicitly contained in the measurement equation as expressed in eq (7.5) and thus far have not been discussed as part of the uncertainty analysis. These are the wavelength uncertainty in the monochromator setting, DVM uncertainty, long-term stability, and the scaling factor for the relative responsivity measurements. Each of these is discussed below and is evaluated in sections 7.2 and 7.3.

<u>Wavelength Uncertainty</u>

The wavelength uncertainty is the uncertainty in the measured spectral responsivity due to the uncertainty of the set wavelength of the monochromator. The repeatability of that wavelength setting is negligible compared to the wavelength uncertainty and is considered part of the relative measurement uncertainty. The wavelength uncertainty of the monochromator $u(\lambda)$ introduces an uncertainty proportional to the slope of the responsivity $dS_x/d\lambda$ [53]

$$u(S_\lambda) = \frac{dS_\lambda}{d\lambda} \cdot u(\lambda) = \frac{d}{d\lambda}\left[\frac{R_{ws}}{R_s} \cdot G \cdot S_s\right] \cdot u(\lambda) \; [\text{U} \cdot \text{W}^{-1}], \text{ where} \quad (7.9)$$

$$S_\lambda = S_x = \frac{R_x}{R_s} \cdot G \cdot S_s \; [\text{U} \cdot \text{W}^{-1}], \quad (7.10)$$

and $G = G_s/G_x$ is the ratio of the amplifier gains, which is a constant with respect to wavelength. Carrying out the differentiation gives

$$u(S_\lambda) = \left[\frac{S_s}{R_s^2} \cdot \left(R_s \cdot \frac{dR_x}{d\lambda} - R_x \cdot \frac{dR_s}{d\lambda}\right) + \frac{R_x}{R_s} \cdot \frac{dS_s}{d\lambda}\right] \cdot u(\lambda) \; [\text{U} \cdot \text{W}^{-1}]. \quad (7.11)$$

The wavelength uncertainty of the monochromator $u(\lambda)$ is the standard deviation of the residuals to a fit of the wavelength calibration.

The relative uncertainty can be shown to be

$$\frac{u(S_\lambda)}{S_\lambda} = \frac{u(\lambda)}{G \cdot S_s} \cdot \frac{R_s}{R_x} \cdot \frac{dS_x}{d\lambda} = \frac{u(\lambda)}{G} \cdot \left[\frac{1}{R_x} \cdot \frac{dR_x}{d\lambda} - \frac{1}{R_s} \cdot \frac{dR_s}{d\lambda} + \frac{1}{S_s} \cdot \frac{dS_s}{d\lambda} \right] \text{ [unitless]}. \qquad (7.12)$$

Typically the gain ratio G equals unity since the same gain range is used for the same detector type (model). For the same type of detector $R_x = R_s$ and $G = 1$, eq (7.12) reduces to

$$\frac{u(S_\lambda)}{S_\lambda} = \frac{u(\lambda)}{S_s} \cdot \frac{dS_s}{d\lambda} \text{ [unitless]}. \qquad (7.13)$$

While the wavelength uncertainty typically is not a large contribution to the measurement uncertainty, it can be significant if the responsivity slope is large. This is especially noticeable with filter and detector packages where the responsivity curve can be very steep in the cut-on and cut-off regions of the filter. A good example is a photometer.

<u>DVM Uncertainty</u>

The values R_{ws} and R_s are the ratio of the difference of four signals which were all taken by DVMs. It follows that the measurement uncertainty of the DVMs are to be included. An uncertainty is calculated for each of the eight signal measurements taken with a DVM (Type B uncertainties) and combined by RSS (ignoring the covariances due to using two DVMs) to give the overly conservative DVM relative uncertainty, $u(V)/V$.

<u>Long-term Stability</u>

The long-term stability is a component of the uncertainty analysis. The relative uncertainty $u(S_{lt})/S_{lt}$ is the relative difference of the measured responsivity over approximately a one year time period. The long-term stability numbers presented in the uncertainty tables could be influenced by temperature changes. For most of the spectral region the temperature effect is less than the stated long-term uncertainty. In the near-IR spectral region (> 960 nm for the Hamamatsu S1337) the temperature effect may be larger than the stated long-term uncertainty. A more thorough investigation could increase these numbers. However, using recently acquired temperature-controlled fixtures for the silicon working standards is expected to reduce the temperature effect significantly, and could reduce the uncertainty component for long-term stability in this spectral region.

The long-term changes in the germanium working standards are not understood, but could be due to reflections from the wedged window reflecting off of the baffle or incident on the monitor detector. This is being studied.

Scaling factor

The scaling factor γ is the ratio of absolute responsivity and relative responsivity measurements. The scaling factor enters the measurement equation with the relative responsivity s_p of the pyroelectric detector. The scaling factor relative uncertainty $u(\gamma)/\gamma$ is the relative standard deviation of the mean of the γ_i's calculated in the spectral region of i wavelengths where the measured responsivities overlap.

7.1.2 Uncertainty Components due to Assumptions and Approximations

Stray Light

Stray light does not add error to measurements when the test and standard detectors have identical responsivity curves. But when the test and standard detectors have different responsivity curves then the error introduced by stray light must be taken into account. This is often very difficult to do, since the slit-scattering function is needed and is very difficult to measure practically; thus it has to be estimated or measured indirectly [54] over the entire spectral range of the monochromator or the responsivity spectral range of the detectors used in operation.

In principle, the stray light is very small when using a double monochromator. Typically this is verified by measurements with various cut-off filters, and the stray light is considered to be a negligible source of error. For this uncertainty analysis, an estimation of the stray light was calculated; and its contribution to the measurement uncertainty was shown to be negligible for almost all of the measurements.

The error due to stray light can be estimated by letting $V_x(\lambda_0)$ be the signal (in U units) from a detector x when the monochromator is set to wavelength λ_0

$$V_x(\lambda_0) = V_{x,bw}(\lambda_0) + V_{x,sl}(\lambda_0) \ [U], \tag{7.14}$$

where $V_{x,bw}(\lambda_0)$ is the signal due to light within the full-width bandpass $\Delta\lambda$ and $V_{x,sl}(\lambda_0)$ is the signal due to stray light. Using eqs (3.3) and (3.4), the measurement equation for the detector signal due to the light within the bandpass $\Delta\lambda$ is

$$V_{x,bw}(\lambda_0) = \int_{\Delta\lambda} S_x(\lambda) \cdot \Phi_{D,\lambda}(\lambda, \lambda_0) \cdot d\lambda \ [U]. \tag{7.15}$$

Similarly, the signal due to stray light is

$$V_{x,sl}(\lambda_0) = \int_{\lambda \neq \Delta\lambda} S_x(\lambda) \cdot \Phi_{D,\lambda}(\lambda, \lambda_0) \cdot d\lambda \ [U], \tag{7.16}$$

where the integral spans the detector responsivity spectral range but is not inclusive of the bandpass $\Delta\lambda$. The incident flux $\Phi_{D,\lambda}(\lambda,\lambda_0)$ on the detector from the monochromator can be expressed as a product with the slit-scattering function

$$\Phi_{D,\lambda}(\lambda,\lambda_0) = z(\lambda_0 - \lambda) \cdot \Phi_{f,\lambda}(\lambda) \; [W], \qquad (7.17)$$

where $z(\lambda_0 - \lambda)$ is the same slit-scattering function introduced in eq (3.6), and $\Phi_{f,\lambda}(\lambda)$ is the spectral radiant flux on the detector at λ_0. (Note: the term $\Phi_{f,\lambda}(\lambda)$, in this situation, implicitly includes the transmittance of the optics and atmosphere unlike in sec. 3.1.) The measurement equation for the detector at wavelength λ_0 is thus

$$V_x(\lambda_0) = \int_{\Delta\lambda} S_x(\lambda) \cdot z(\lambda_0 - \lambda) \cdot \Phi_{f,\lambda}(\lambda) \cdot d\lambda + \int_{\lambda \neq \Delta\lambda} S_x(\lambda) \cdot z(\lambda_0 - \lambda) \cdot \Phi_{f,\lambda}(\lambda) \cdot d\lambda \; [U]. \qquad (7.18)$$

Similarly, the measurement equation for a working standard detector s is

$$V_s(\lambda_0) = \int_{\Delta\lambda} S_s(\lambda) \cdot z(\lambda_0 - \lambda) \cdot \Phi_{f,\lambda}(\lambda) \cdot d\lambda + \int_{\lambda \neq \Delta\lambda} S_s(\lambda) \cdot z(\lambda_0 - \lambda) \cdot \Phi_{f,\lambda}(\lambda) \cdot d\lambda \; [U]. \qquad (7.19)$$

For estimating the stray light contribution to the measured signal, the slit-scattering function is assumed to have the computationally convenient form

$$z(\lambda_0 - \lambda) = \begin{cases} 1 & \text{for } \lambda_0 - \Delta\lambda \leq \lambda \leq \lambda_0 + \Delta\lambda \\ 10^{-7} & \text{otherwise} \end{cases}. \qquad (7.20)$$

The error due to stray light is estimated by converting the integrals to summations and taking the difference between the ratio of the measured signals from the test and working standard detectors, R_T, and the ratio R_{bw} of the signals due to light within the bandpass $\Delta\lambda$. The relative error due to stray light when the monochromator is set to λ_0 is estimated as

$$\frac{\delta R_{sl}(\lambda_0)}{R_{sl}(\lambda_0)} \approx \frac{R_T(\lambda_0) - R_{bw}(\lambda_0)}{R_{bw}(\lambda_0)} \; [\text{unitless}], \qquad (7.21)$$

where $\qquad R_T(\lambda_0) = \dfrac{V_x(\lambda,\lambda_0)}{V_s(\lambda,\lambda_0)}$ and $R_{sl}(\lambda_0) = R_{bw}(\lambda_0) = \dfrac{V_{x,bw}(\lambda,\lambda_0)}{V_{sbw}(\lambda,\lambda_0)}$. $\qquad$ (7.22a,b)

The relative error was calculated over the measurement spectral range at 5 nm intervals of λ_0 and will be shown below. Because these errors are small, a correction is not applied, but instead the component of relative standard uncertainty is assumed equal to the relative error. It can be seen from tables 7.1 to 7.7 that this component makes a negligible contribution to the combined standard uncertainty.

Bandwidth-effect

As mentioned in section 3.2.4, the bandwidth of the monochromator output flux causes an error in the measurement when the monochromator wavelength scale is calibrated using the peak wavelength of the bandpass and the responsivity curves of the test and standard detectors do not have the same slope. This error can be approximated for the uncertainty analysis. Consider the case of a pyroelectric detector used as a standard. The responsivity for the standard is assumed constant (flat) and the responsivities for the test and monitor detector are assumed linear over the bandpass. This is the situation when measuring the relative responsivity of working standard photodiodes. Following the development of eq (17) in Ref. [53] it can be shown that the measurement equation for spectral responsivity, taking into consideration the bandwidth of the monochromator,

$$S_x = \frac{R_x}{R_s} \cdot S_s - \frac{b \cdot \beta^2}{6} \cdot \frac{dS_x}{d\lambda} \quad [\text{U} \cdot \text{W}^{-1}], \tag{7.23}$$

where the ratio of signals is now replaced by the ratio of test to monitor and standard to monitor signal ratios. Note the change in responsivity and bandwidth notation from Ref. [53]. The other terms are defined as $2\beta = \Delta\lambda$ = full-width bandpass ($\beta = \Delta\lambda/2$ = FWHM for triangular bandpass)

$$b = \frac{s}{\Phi_D - \lambda_0 \cdot s} \quad [\text{m}^{-1}], \text{ where } s = \frac{d\Phi_D}{d\lambda} \quad [\text{W} \cdot \text{m}^{-1}], \tag{7.24a,b}$$

and $d\Phi_D$ is the change in flux reaching the detector (i.e., s = slope of the output flux function). This assumes a triangular bandpass for the monochromator and that the flux varies linearly over the bandpass.

When the responsivity of the standard also changes spectrally, it can be shown that the measurement equation is similar to eq (18) in Ref. [53],

$$S_x = \frac{R_x}{R_s} \cdot \left[S_s + \frac{b \cdot \beta^2}{6} \cdot \frac{dS_s}{d\lambda} \right] - \frac{b \cdot \beta^2}{6} \cdot \frac{dS_x}{d\lambda} \quad [\text{U} \cdot \text{W}^{-1}], \tag{7.25}$$

where all the terms and conditions are the same as before. This is the general test situation. The bandwidth-effect error component, using eq (7.23), is

$$\delta S_{bw} = C_{bw} = -\frac{b \cdot \beta^2}{6} \cdot \frac{dS_x}{d\lambda} \quad [\text{U} \cdot \text{W}^{-1}]. \tag{7.26}$$

The relative error is

$$\frac{\delta S_{bw}}{S_{bw}} = \frac{\delta S_{bw}}{\frac{R_x}{R_s} \cdot S_s} \quad [\text{unitless}], \tag{7.27}$$

and

$$\frac{\delta S_{bw}}{S_{bw}} = \frac{\delta S_{bw}}{S_s} \cdot \frac{R_s}{R_x} = \frac{-b \cdot \beta^2}{6 \cdot S_s} \cdot \frac{R_s}{R_x} \cdot \frac{\delta_x S}{d\lambda} \text{ [unitless]}. \quad (7.28)$$

Similarly, for eq (7.25) the relative error is eq (7.27) where the bandwidth-effect uncertainty component is

$$\delta S_{bw} = C_{bw} = \frac{R_x}{R_s} \cdot \frac{b \cdot \beta^2}{6} \cdot \frac{dS_s}{d\lambda} - \frac{b \cdot \beta^2}{6} \cdot \frac{dS_x}{d\lambda} \text{ [U·W}^{-1}\text{]}, \quad (7.29)$$

and

$$\frac{\delta S_{bw}}{S_{bw}} = \frac{\delta S_{bw}}{S_s} \cdot \frac{R_s}{R_x} = \frac{b \cdot \beta^2}{6 \cdot S_s} \cdot \left[\frac{dS_s}{d\lambda} - \frac{R_s}{R_x} \cdot \frac{dS_x}{d\lambda}\right] \text{ [unitless]}. \quad (7.30)$$

As stated in Ref. [53] and in section 3.2.4, the effect of the bandwidth could be treated as a correction term C_{bw}, but since it is small for the measurements described here it is applied as an uncertainty component. Numerical results are presented in tables 7.1 to 7.7 and it can be seen that this component makes a negligible contribution to the combined standard uncertainty.

7.1.3 Other Factors Considered and Neglected

Several other factors that could contribute uncertainty components were mentioned in section 3.1.1 and are discussed here. Stable detector responsivities are fundamental for detector-based radiometry. Detector responsivity stability has been the subject of several past and present studies reported [55]; and, for this application, short-term instability has not been observed. Long-term stability has already been discussed. Effects of polarization have been studied and found to be negligible for the typical situation, where the detectors are measured at normal incidence to the optical axis (and the detector surfaces are isotropic). The effect of the converging beam angle on the reflectance (and transmittance) from the detector surface (and window) is small compared to the variance of repeated measurements and is typically neglected. Water condensation (onto the detector or window) and the effects of water absorption have not been observed and are neglected.

The responsivity temperature coefficient can be significant, especially for wavelengths near the bandgap of a photodiode. Typically, the temperature variation is small over the measurement time and is accounted for in the relative measurement uncertainty since the measurement consists of multiple scans. Temperature variation was considered in the long-term stability estimation. Therefore no explicit term is included in the uncertainty estimate for the responsivity temperature dependence. As mentioned earlier, diffraction and coherence effects are negligible for these comparator systems.

The responsivity spatial uniformity is typically measured but not explicitly used in the uncertainty analysis. Instead, like the temperature variations, it is considered part of the relative measurement uncertainty and long-term stability uncertainty components since the spatial uniformity directly affects these measurements (i.e., the reproducibility). The same holds true for the monochromator beam profile and shape.

Geometrically scattered (stray) light is light scattered out of the nominal monochromator beam and is thought to be primarily due to imperfections in the optics. The geometrically scattered light cancels when the test and standard detectors are large enough to collect all of the radiation from the monochromator over an area of uniform responsivity. This is the typical measurement condition and thus the geometrically scattered light is neglected. Background radiation (sometimes called diffuse stray light) has been measured and found to be negligible.

The reflected beam from the test detector(s) and working standard detectors travels "backwards" along the optical axis and assumed to be scattered inside the monochromator. Any of this reflected beam that returns to the detector is considered as part of the geometrically scattered light discussed above. The reflected beam from the monitor is not returned along the same path and is diffusely scattered inside the enclosure on the opposite side of the baffle from the test detector and working standard detectors. It contributes to the background radiation previously mentioned.

Effects of detector nonlinearity have been discussed in the literature [32, 34, 56]. The detector linearity has been measured for the Hamamatsu S1337-1010BQ and the UDT Sensors UV100 silicon photodiodes and is discussed in section 9. At the power levels used for routine measurements, nonlinearity is not a consideration.

The frequency response of the detector and amplifier is important to consider when comparing absolute measurements between chopped (optically modulated or ac) and dc measurements. But since the chopped (pyroelectric comparison) measurements are relative measurements, corrections due to frequency response effects are not considered.

7.2 Transfer from Traps (HACR) to Working Standards

In this section, a detailed listing of the uncertainty components for each type of working standard is provided along with a description of how each component was obtained. The data was calculated at 5 nm intervals as shown in the figures; abbreviated numerical tables are provided. The relative measurement uncertainty was determined by calculating the RSS of the weighted uncertainties from multiple measurements. The amplifier calibration uncertainties are all identical because identical amplifiers are used and they are calibrated using the same equipment and procedure. Stray light, bandwidth-effect, and wavelength calibration uncertainties were all calculated as described above using typical data. All components were considered independent and combined by RSS to get the relative combined standard uncertainty shown.

7.2.1 Visible Silicon Working Standards

Table 7.1 lists the uncertainty components for the three calibrations used for the visible silicon working standards. The UV and Ge working standard calibration uncertainties are the combined standard uncertainties from tables 7.3 and 7.2, respectively. The trap uncertainties are from the HACR transfer [21]. The long-term stability is the typical data for a visible working standard over a one year period. All the components were combined by RSS to get the relative combined standard uncertainty shown at 5 nm intervals in figure 7.1.

Table 7.1. Visible Working Standard Uncertainty Transfer from Traps, UV WS, and Ge WS

Source of uncertainty	Relative measurement uncertainty	DVM uncertainty	UV WS calibration	UV WS amplifier gain	Visible WS amplifier gain	Wavelength calibration (± 0.1 nm)	Stray light	Bandwidth-effect	Visible WS long-term stability	Relative combined standard uncertainty
Type	A	B	B	B	B	B	B	B	A	[%]
Relative uncertainty	$u(R_m)/R_m$	$u(V)/V$	$u(S_{UV})/S_{UV}$	$u(G_{UV})/G_{UV}$	$u(G_{Vis})/G_{Vis}$	$u(S_\lambda)/S_\lambda$	$u(R_{sl})/R_{sl}$	$u(S_{bw})/S_{bw}$	$u(S_{lt})/S_{lt}$	$u_c(S_{Vis})/S_{Vis}$
Wavelength [nm]				Estimated value [%]						Root-sum-of-squares
350	0.87	0.56	0.83	0.04	0.04	0.04	0.000	-0.001	0.06	**1.33**
400	0.23	0.12	0.71	0.04	0.04	-0.03	0.000	0.001	0.03	**0.76**

Source of uncertainty	Relative measurement uncertainty	DVM uncertainty	Trap detector calibration	Trap detector amplifier gain	Visible WS amplifier gain	Wavelength calibration (± 0.1 nm)	Stray light	Bandwidth-effect	Visible WS long-term stability	Relative combined standard uncertainty
Type	A	B	B	B	B	B	B	B	A	[%]
Relative uncertainty	$u(R_m)/R_m$	$u(V)/V$	$u(S_T)/S_T$	$u(G_T)/G_T$	$u(G_{Vis})/G_{Vis}$	$u(S_\lambda)/S_\lambda$	$u(R_{sl})/R_{sl}$	$u(S_{bw})/S_{bw}$	$u(S_{lt})/S_{lt}$	$u_c(S_{Vis})/S_{Vis}$
Wavelength [nm]				Estimated value [%]						Root-sum-of-squares
405	0.03	0.05	0.05	0.04	0.04	-0.05	0.001	-0.001	0.05	**0.12**
450	0.03	0.02	0.05	0.04	0.04	-0.03	0.000	-0.001	0.01	**0.09**
500	0.03	0.01	0.05	0.04	0.04	-0.02	0.000	0.000	0.01	**0.08**
550	0.03	0.01	0.05	0.04	0.04	-0.02	0.000	0.000	0.01	**0.08**
600	0.02	0.01	0.05	0.04	0.04	-0.02	0.000	0.000	0.01	**0.08**
650	0.03	0.01	0.05	0.04	0.04	-0.02	0.000	0.000	0.01	**0.08**
700	0.03	0.01	0.05	0.04	0.04	-0.01	0.000	0.000	0.01	**0.08**
750	0.03	0.01	0.05	0.04	0.04	-0.01	0.000	0.000	0.02	**0.08**
800	0.02	0.02	0.05	0.04	0.04	-0.01	0.000	0.000	0.03	**0.09**
850	0.02	0.02	0.05	0.04	0.04	-0.01	0.000	0.000	0.04	**0.09**
900	0.02	0.01	0.05	0.04	0.04	-0.01	0.000	0.000	0.04	**0.09**
920	0.02	0.01	0.05	0.04	0.04	-0.01	0.000	0.000	0.05	**0.10**

Source of uncertainty	Relative measurement uncertainty	DVM uncertainty	Ge WS calibration	Ge WS amplifier gain	Visible WS amplifier gain	Wavelength calibration (± 0.1 nm)	Stray light	Bandwidth-effect	Visible WS long-term stability	Relative combined standard uncertainty
Type	A	B	B	B	B	B	B	B	A	[%]
Relative uncertainty	$u(R_m)/R_m$	$u(V)/V$	$u(S_{Ge})/S_{Ge}$	$u(G_{Ge})/G_{Ge}$	$u(G_{Vis})/G_{Vis}$	$u(S_\lambda)/S_\lambda$	$u(R_{sl})/R_{sl}$	$u(S_{bw})/S_{bw}$	$u(S_{lt})/S_{lt}$	$u_c(S_{Vis})/S_{Vis}$
Wavelength [nm]				Estimated value [%]						Root-sum-of-squares
950	0.90	0.02	0.90	0.04	0.04	-0.02	0.002	0.001	0.22	**1.29**
1000	0.39	0.01	0.75	0.04	0.04	0.03	0.001	0.002	0.13	**0.86**
1050	0.43	0.01	0.87	0.04	0.04	0.17	0.000	0.003	0.87	**1.31**
1100	0.40	0.02	0.78	0.04	0.04	0.18	-0.002	0.001	1.86	**2.06**

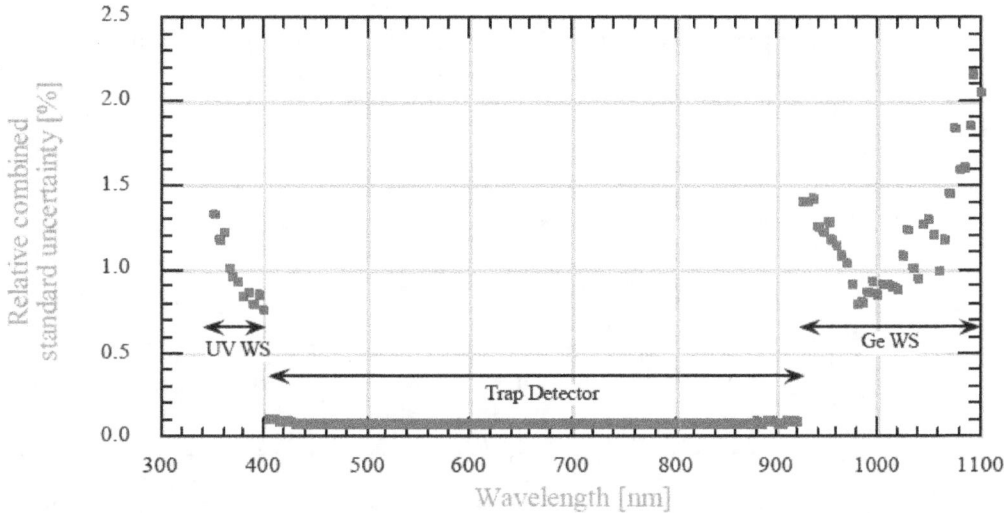

Figure 7.1. Visible Working Standard (Vis WS) relative combined standard uncertainty. The discontinuities in the curve are the result of calibrations with different standards.

7.2.2 Germanium (NIR) Working Standards

Table 7.2 lists the uncertainty components for the relative responsivity measurements using the pyroelectric detector and the absolute responsivity transfer measurement uncertainty components using the HACR-calibrated trap detectors. The absolute measurements are used to scale the relative data of the germanium (NIR) working standards. The low SNR when using the pyroelectric gives rise to larger relative measurement uncertainty values. Also, the long-term stability adds significantly to the germanium working standard uncertainty. The relative combined standard uncertainty was determined by the RSS of all the components. Figure 7.2 shows the relative combined standard uncertainty at 5 nm intervals.

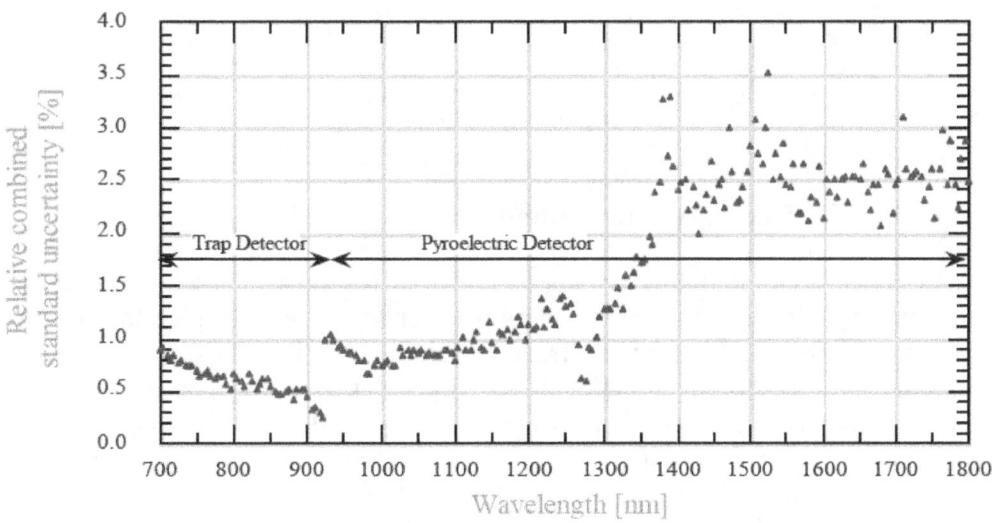

Figure 7.2. Germanium (NIR) Working Standard (Ge WS) relative combined standard uncertainty.

Table 7.2. Germanium Working Standard Uncertainty

Transfer from pyroelectric (relative) and scaling with trap (absolute)

Source of uncertainty	Relative measurement uncertainty	DVM uncertainty	Trap detector calibration	Trap detector amplifier gain	Ge WS amplifier gain	Wavelength calibration (± 0.1 nm)	Stray light	Bandwidth-effect	Ge WS long-term stability	Relative combined standard uncertainty
Type	A	B	B	B	B	B	B	B	A	[%]
Relative uncertainty	$u(R_m)/R_m$	$u(V)/V$	$u(S_T)/S_T$	$u(G_T)/G_T$	$u(G_{Ge})/G_{Ge}$	$u(S_\lambda)/S_\lambda$	$u(R_{sl})/R_{sl}$	$u(S_{bw})/S_{bw}$	$u(S_{lt})/S_{lt}$	$u_c(S_{Ge})/S_{Ge}$
Wavelength [nm]					Estimated value [%]					Root-sum-of-squares
700	0.05	0.03	0.05	0.04	0.04	-0.09	0.008	0.002	0.87	**0.88**
750	0.05	0.03	0.05	0.04	0.04	-0.07	0.008	0.002	0.69	**0.70**
800	0.04	0.04	0.05	0.04	0.04	-0.06	0.008	0.001	0.65	**0.66**
850	0.03	0.04	0.05	0.04	0.04	-0.05	0.008	0.001	0.54	**0.55**
900	0.03	0.02	0.05	0.04	0.04	-0.04	0.003	0.001	0.44	**0.45**
920	0.03	0.02	0.05	0.04	0.04	-0.04	0.003	0.001	0.23	**0.25**

Source of uncertainty	Relative measurement uncertainty	DVM uncertainty	Pyroelectric relative calibration	Scaling factor γ	Wavelength calibration (± 0.1 nm)	Stray light	Bandwidth-effect	Ge WS long-term stability	Relative combined standard uncertainty
Type	A	B	B	B	B	B	B	A	[%]
Relative uncertainty	$u(R_m)/R_m$	$u(V)/V$	$u(s_p)/s_p$	$u(\gamma)/\gamma$	$u(S_\lambda)/S_\lambda$	$u(R_{sl})/R_{sl}$	$u(S_{bw})/S_{bw}$	$u(S_{lt})/S_{lt}$	$u_c(S_{Ge})/S_{Ge}$
Wavelength [nm]				Estimated value [%]					Root-sum-of-squares
950	0.68	0.02	0.52	0.14	-0.02	0.000	0.001	0.26	**0.90**
1000	0.39	0.01	0.52	0.14	-0.06	0.000	0.000	0.35	**0.75**
1050	0.63	0.01	0.52	0.14	-0.01	-0.001	0.000	0.25	**0.87**
1100	0.56	0.01	0.52	0.14	0.00	-0.001	0.000	0.09	**0.78**
1150	0.78	0.01	0.52	0.14	0.00	-0.001	-0.001	0.15	**0.96**
1200	0.98	0.01	0.52	0.14	0.00	-0.001	0.000	0.18	**1.13**
1250	1.17	0.01	0.52	0.14	-0.03	-0.001	0.000	0.11	**1.30**
1300	1.14	0.01	0.52	0.14	-0.01	-0.001	0.000	0.21	**1.28**
1350	1.64	0.01	0.52	0.14	-0.05	-0.002	0.000	-0.05	**1.73**
1400	2.34	0.02	0.52	0.14	0.02	-0.003	0.000	-0.04	**2.40**
1450	2.25	0.01	0.52	0.14	0.07	-0.003	0.000	-0.11	**2.31**
1500	2.77	0.01	0.52	0.14	0.06	-0.003	0.000	-0.11	**2.83**
1550	2.34	0.02	0.52	0.14	0.08	-0.003	-0.001	0.57	**2.46**
1600	1.47	0.02	0.52	0.14	-0.02	-0.002	0.000	1.47	**2.14**
1650	1.75	0.03	0.52	0.14	0.08	0.000	0.000	1.72	**2.52**
1700	1.69	0.05	0.52	0.14	0.04	0.003	-0.001	1.70	**2.46**
1750	2.10	0.09	0.52	0.14	0.18	0.016	-0.001	1.45	**2.61**
1800	2.12	0.26	0.52	0.14	0.06	0.064	-0.001	1.15	**2.49**

7.2.3 UV Silicon Working Standards

Table 7.3 lists the uncertainty components for the relative responsivity measurements using the pyroelectric detector and the absolute responsivity transfer measurement uncertainty components from the visible working standards which are traceable to the HACR. The absolute measurements are used to scale the relative data of the UV silicon working standards. The pyroelectric relative measurement uncertainty was affected by a low SNR below 250 nm. The long-term stability is also a large component in the uncertainty. As with the previous working standards, the relative combined standard uncertainty was determined by the RSS of each component. Figure 7.3 shows the relative combined standard uncertainty at 5 nm intervals.

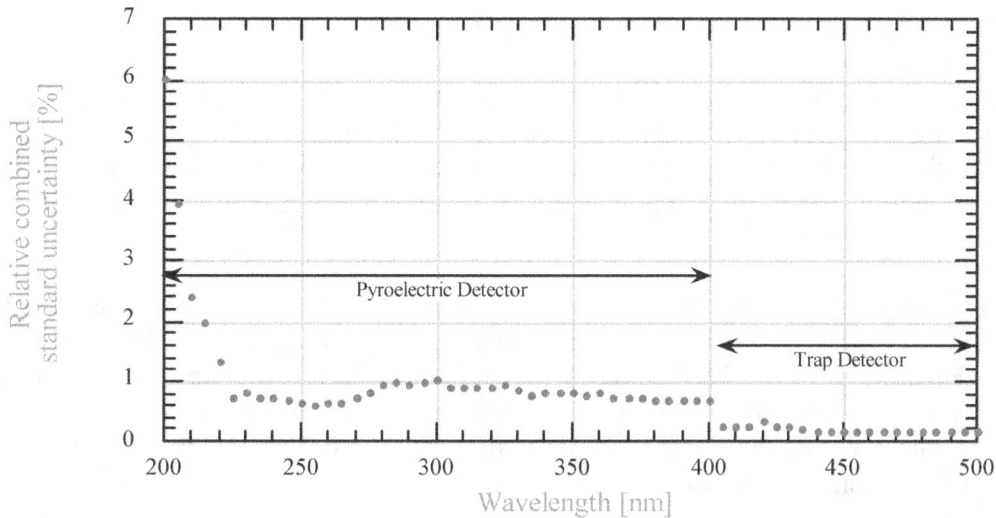

Figure 7.3. Ultraviolet Working Standard (UV WS) relative combined standard uncertainty.

Table 7.3. UV Working Standard Uncertainty

Transfer from Pyroelectric (relative) and Scaling with Visible WS (absolute)

Source of uncertainty	Relative measurement uncertainty	DVM uncertainty	Pyroelectric relative calibration	Scaling factor γ	Wavelength calibration ($\pm$ 0.1 nm)	Stray light	Bandwidth-effect	UV WS long-term stability	Relative combined standard uncertainty [%]
Type	A	B	B	B	B	B	B	A	
Relative uncertainty	$u(R_m)/R_m$	$u(V)/V$	$u(s_p)/s_p$	$u(\gamma)/\gamma$	$u(S_\lambda)/S_\lambda$	$u(R_{sl})/R_{sl}$	$u(S_{bw})/S_{bw}$	$u(S_{lt})/S_{lt}$	$u_c(S_{UV})/S_{UV}$
Wavelength [nm]				Estimated value [%]					Root-sum-of-squares
200	5.87	0.38	0.52	0.18	-0.98	0.160	0.002	0.72	**6.03**
250	0.34	0.01	0.52	0.18	-0.05	0.001	0.001	-0.19	**0.67**
300	0.38	0.01	0.52	0.18	-0.03	0.000	0.001	0.77	**1.02**
350	0.44	0.01	0.52	0.18	0.03	0.001	0.000	0.44	**0.83**
400	0.42	0.01	0.52	0.18	-0.06	0.000	0.000	0.17	**0.71**

Source of uncertainty	Relative measurement uncertainty	DVM uncertainty	Visible WS calibration	Visible WS amplifier gain	UV WS amplifier gain	Wavelength calibration ($\pm$ 0.1 nm)	Stray light	Bandwidth-effect	UV WS long-term stability	Relative combined standard uncertainty [%]
Type	A	B	B	B	B	B	B	B	A	
Relative uncertainty	$u(R_m)/R_m$	$u(V)/V$	$u(S_{Vis})/S_{Vis}$	$u(G_{Vis})/G_{Vis}$	$u(G_{UV})/G_{UV}$	$u(S_\lambda)/S_\lambda$	$u(R_{sl})/R_{sl}$	$u(S_{bw})/S_{bw}$	$u(S_{lt})/S_{lt}$	$u_c(S_{UV})/S_{UV}$
Wavelength [nm]					Estimated value [%]					Root-sum-of-squares
450	0.12	0.04	0.09	0.04	0.04	-0.05	0.0000	-0.002	0.00	**0.17**
500	0.11	0.03	0.08	0.04	0.04	-0.04	-0.0001	-0.001	-0.04	**0.16**

7.3 Transfer to Test (Customer) Detectors

This section details the uncertainty components when transferring the spectral responsivity scale from the working standards to test (customer) detectors. The tables and figures are calculated for specific photodiode models and would, in general, be different for other detectors. These tables and figures serve as a starting point to estimate the minimum uncertainty value possible from this measurement service using the current configuration.

The uncertainty in the spectral responsivity transferred to a test detector is determined similarly to the working standards using eq (7.8). Since, in general, it is not possible to know the details of how a test detector is used, NIST policy [7] is to not include estimates of the uncertainties introduced by transporting the detector or its use by the customer as a reference standard. Examples are uncertainties due to the passage of time (long-term stability) and differences in laboratory environmental conditions.

The stray light and bandwidth-effect are negligible and effects due to wavelength calibration uncertainty are minimized when the working standard and test detector have the same response curve. In the case of an InGaAs photodiode calibrated with the germanium working standards, stray light and bandwidth-effect are almost completely negligible except near the ends of the responsivity spectral range.

As with the transfer to the working standards, abbreviated numerical tables are provided even though the data was calculated at 5 nm intervals; full data sets are shown in the figures. The relative measurement uncertainty was determined by the RSS of the weighted uncertainties from multiple measurements except for the InGaAs example where only one measurement was made and the weighted uncertainties were used. The amplifier calibration uncertainties are all identical because the amplifiers used are identical and calibrated in the same way. Bandwidth-effects, stray light, and wavelength calibration uncertainties were all calculated using typical data as described above. All uncertainty components were considered independent and were combined by RSS to get the relative combined standard uncertainty shown.

Figure 7.4 shows the relative combined standard uncertainty at 5 nm intervals for three types of transfer measurements routinely provided to customers. Note that the NIST policy is to report to customers the uncertainty using a expansion factor of $k = 2$. The expanded uncertainties are shown in figure 2.1.

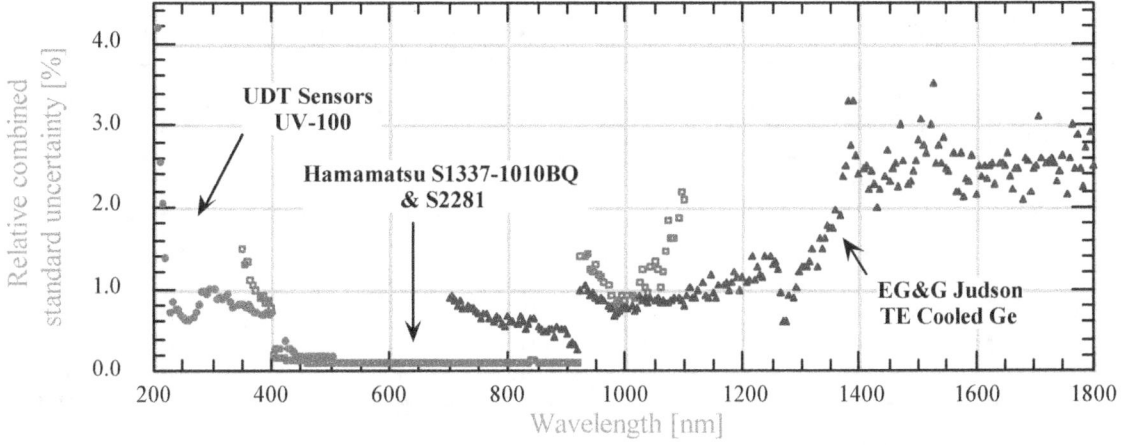

(Note: Relative standard uncertainty at 200 nm is 6.5 %.)

Figure 7.4. Transfer to test (customer) detectors relative combined standard uncertainty.

7.3.1 UV Silicon Transfer

Table 7.4 lists the uncertainty components for the transfer comparison measurements for test UV silicon photodiodes using the UV working standards. The UV working standard calibration uncertainties are the combined standard uncertainties from table 7.3. The transfer relative combined standard uncertainty was determined by the RSS of the components.

Table 7.4. Transfer Uncertainty to Test (Customer) UV100 Silicon Photodiodes

Source of uncertainty	Relative measurement uncertainty	DVM uncertainty	UV WS calibration	UV WS amplifier gain	Test detector amplifier gain	Wavelength calibration (± 0.1 nm)	Stray light[†]	Bandwidth-effect[†]	Relative combined standard uncertainty
Type	A	B	B	B	B	B	B	B	[%]
Relative uncertainty	$u(R_m)/R_m$	$u(V)/V$	$u(S_{UV})/S_{UV}$	$u(G_{UV})/G_{UV}$	$u(G_{Test})/G_{Test}$	$u(S_\lambda)/S_\lambda$	$u(R_{sl})/R_{sl}$	$u(S_{bw})/S_{bw}$	$u_c(S_{Test})/S_{Test}$
Wavelength [nm]				Estimated value [%]					Root-sum-of-squares
200	2.41	0.38	6.03	0.04	0.04	-0.05	0.00	0.00	**6.51**
250	0.09	0.01	0.67	0.04	0.04	-0.05	0.00	0.00	**0.68**
300	0.10	0.01	1.02	0.04	0.04	-0.03	0.00	0.00	**1.03**
350	0.09	0.01	0.83	0.04	0.04	0.03	0.00	0.00	**0.84**
400	0.15	0.01	0.71	0.04	0.04	-0.06	0.00	0.00	**0.73**
450	0.04	0.01	0.17	0.04	0.04	-0.05	0.00	0.00	**0.19**
500	0.05	0.01	0.16	0.04	0.04	-0.04	0.00	0.00	**0.19**

[†]These terms are zero for identical responsivities.

7.3.2 Visible Silicon Transfer

Table 7.5 lists the uncertainty components for the transfer comparison measurements for test Hamamatsu S1337 and S2281 visible silicon photodiodes using the visible working standards. The visible working standard calibration uncertainties are the combined standard uncertainties from table 7.1. The transfer relative combined standard uncertainty was determined by the RSS of the components.

Table 7.5. Transfer Uncertainty to Test (Customer) S1337 and S2281 Silicon Photodiodes

Source of uncertainty	Relative measurement uncertainty	DVM uncertainty	Visible WS calibration	Visible WS amplifier gain	Test detector amplifier gain	Wavelength calibration (± 0.1 nm)	Stray light[†]	Bandwidth-effect[†]	Relative combined standard uncertainty [%]
Type	A	B	B	B	B	B	B	B	[%]
Relative uncertainty	$u(R_m)/R_m$	$u(V)/V$	$u(S_{Vis})/S_{Vis}$	$u(G_{Vis})/G_{Vis}$	$u(G_{Test})/G_{Test}$	$u(S_\lambda)/S_\lambda$	$u(R_{sl})/R_{sl}$	$u(S_{bw})/S_{bw}$	$u_c(S_{Test})/S_{Test}$
Wavelength [nm]				Estimated value [%]					Root-sum-of-squares
350	0.31	0.56	1.33	0.04	0.04	0.01	0.00	0.00	**1.48**
400	0.05	0.12	0.76	0.04	0.04	-0.05	0.00	0.00	**0.78**
450	0.02	0.04	0.09	0.04	0.04	-0.03	0.00	0.00	**0.12**
500	0.01	0.02	0.08	0.04	0.04	-0.02	0.00	0.00	**0.11**
550	0.00	0.02	0.08	0.04	0.04	-0.02	0.00	0.00	**0.10**
600	0.01	0.01	0.08	0.04	0.04	-0.02	0.00	0.00	**0.10**
650	0.01	0.01	0.08	0.04	0.04	-0.02	0.00	0.00	**0.10**
700	0.01	0.02	0.08	0.04	0.04	-0.01	0.00	0.00	**0.10**
750	0.01	0.02	0.08	0.04	0.04	-0.01	0.00	0.00	**0.11**
800	0.02	0.03	0.09	0.04	0.04	-0.01	0.00	0.00	**0.11**
850	0.02	0.03	0.09	0.04	0.04	-0.01	0.00	0.00	**0.11**
900	0.01	0.02	0.09	0.04	0.04	-0.01	0.00	0.00	**0.11**
950	0.01	0.01	1.29	0.04	0.04	-0.01	0.00	0.00	**1.29**
1000	0.01	0.01	0.86	0.04	0.04	0.04	0.00	0.00	**0.86**
1050	0.04	0.02	1.31	0.04	0.04	0.18	0.00	0.00	**1.33**
1100	0.08	0.03	2.06	0.04	0.04	0.23	0.00	0.00	**2.08**

[†]These terms are zero for identical responsivities.

7.3.3 NIR Transfer

The transfer comparison measurement uncertainties for example germanium and indium gallium arsenide (InGaAs) test photodiodes are presented in tables 7.6 and 7.7 respectively. Several of the data columns are detector-dependent, thus these tables are presented as references of typical uncertainties for these types of photodiodes. The reported uncertainties do not include estimates for several components that are unknown for these test photodiodes. The unknown uncertainty components are photodiode responsivity uniformity, polarization sensitivity, linearity, temperature coefficient, and long-term stability. These components could significantly add to the reported uncertainty.

7.3.3.1 Germanium Transfer

Table 7.6 lists the uncertainty components for the transfer comparison measurements for test thermoelectrically cooled germanium photodiodes using the germanium working standards. The germanium NIR working standard calibration uncertainties are the combined standard uncertainties from table 7.2. The transfer relative combined standard uncertainty was determined by the RSS of the components.

The transfer relative combined standard uncertainty presented in table 7.6 is for a Judson EG&G J16TE2-8A6-R05M-SC test photodiode. For different customer-supplied germanium photodiodes the uncertainty is reanalyzed for that device.

Table 7.6. Transfer Uncertainty to Test (Customer) TE Cooled Germanium Photodiodes

Source of uncertainty	Relative measurement uncertainty	DVM uncertainty	Ge WS calibration	Ge WS amplifier gain	Test detector amplifier gain	Wavelength calibration (± 0.1 nm)	Stray light[†]	Bandwidth-effect[†]	Relative combined standard uncertainty [%]
Type	A	B	B	B	B	B	B	B	
Relative uncertainty	$u(R_m)/R_m$	$u(V)/V$	$u(S_{Ge})/S_{Ge}$	$u(G_{Ge})/G_{Ge}$	$u(G_{Test})/G_{Test}$	$u(S_\lambda)/S_\lambda$	$u(R_{sl})/R_{sl}$	$u(S_{bw})/S_{bw}$	$u_c(S_{Test})/S_{Test}$
Wavelength [nm]				Estimated value [%]					Root-sum-of-squares
700	0.01	0.05	0.88	0.04	0.04	-0.09	0.00	0.00	0.89
750	0.01	0.05	0.70	0.04	0.04	-0.07	0.00	0.00	0.71
800	0.03	0.06	0.66	0.04	0.04	-0.06	0.00	0.00	0.66
850	0.03	0.06	0.55	0.04	0.04	-0.05	0.00	0.00	0.56
900	0.02	0.03	0.45	0.04	0.04	-0.04	0.00	0.00	0.46
950	0.02	0.02	0.90	0.04	0.04	-0.03	0.00	0.00	0.90
1000	0.01	0.01	0.75	0.04	0.04	-0.03	0.00	0.00	0.75
1050	0.01	0.01	0.87	0.04	0.04	-0.03	0.00	0.00	0.87
1100	0.01	0.01	0.78	0.04	0.04	-0.02	0.00	0.00	0.78
1150	0.01	0.01	0.96	0.04	0.04	-0.02	0.00	0.00	0.96
1200	0.01	0.01	1.13	0.04	0.04	-0.02	0.00	0.00	1.13
1250	0.01	0.01	1.30	0.04	0.04	-0.01	0.00	0.00	1.30
1300	0.01	0.01	1.28	0.04	0.04	-0.01	0.00	0.00	1.28
1350	0.02	0.01	1.73	0.04	0.04	-0.01	0.00	0.00	1.73
1400	0.03	0.02	2.40	0.04	0.04	-0.01	0.00	0.00	2.40
1450	0.02	0.01	2.31	0.04	0.04	-0.01	0.00	0.00	2.32
1500	0.03	0.01	2.83	0.04	0.04	-0.01	0.00	0.00	2.83
1550	0.03	0.02	2.46	0.04	0.04	0.11	0.00	0.00	2.47
1600	0.03	0.02	2.14	0.04	0.04	0.06	0.00	0.00	2.15
1650	0.03	0.03	2.52	0.04	0.04	0.06	0.00	0.00	2.52
1700	0.03	0.05	2.46	0.04	0.04	0.10	0.00	0.00	2.46
1750	0.03	0.09	2.61	0.04	0.04	0.18	0.00	0.00	2.62
1800	0.04	0.26	2.49	0.04	0.04	0.20	0.00	0.00	2.51

[†]These terms are zero for identical responsivities.

7.3.3.2 InGaAs Transfer

Table 7.7 lists the uncertainty components for the transfer comparison measurements for test InGaAs photodiodes using the germanium working standards. The weighted uncertainty of one measurement (of three scans) was used as the relative measurement uncertainty. The germanium (NIR) working standard calibration uncertainties are the combined standard uncertainties from table 7.2. The transfer relative combined standard uncertainty was determined by the RSS of the components.

The transfer relative combined standard uncertainty presented in table 7.7 is for an example InGaAs test photodiode. The InGaAs transfer uncertainties are very similar to the germanium transfer uncertainties and were omitted from figure 7.4 for clarity. As with the germanium transfer uncertainty in table 7.6 several of the data columns are detector-dependent, thus this table is presented as a reference representing typical uncertainties for this type of photodiode. For customer-supplied photodiodes the uncertainty is reanalyzed for that device.

Table 7.7. Transfer Uncertainty to Test (Customer) InGaAs Photodiodes

Source of uncertainty	Relative measurement uncertainty	DVM uncertainty	Ge WS calibration	Ge WS amplifier gain	Test detector amplifier gain	Wavelength calibration (± 0.1 nm)	Stray light	Bandwidth-effect	Relative combined standard uncertainty
Type	A	B	B	B	B	B	B	B	[%]
Relative uncertainty	$u(R_m)/R_m$	$u(V)/V$	$u(S_{Ge})/S_{Ge}$	$u(G_{Ge})/G_{Ge}$	$u(G_{Test})/G_{Test}$	$u(S_\lambda)/S_\lambda$	$u(R_{sl})/R_{sl}$	$u(S_{bw})/S_{bw}$	$u_c(S_{Test})/S_{Test}$
Wavelength [nm]				Estimated value [%]					Root-sum-of-squares
700	0.16	0.04	0.88	0.04	0.04	-0.06	0.01	0.00	**0.90**
750	0.18	0.05	0.70	0.04	0.04	-0.06	0.01	0.00	**0.73**
800	0.17	0.06	0.66	0.04	0.04	-0.06	0.01	0.00	**0.69**
850	0.13	0.06	0.55	0.04	0.04	-0.11	0.01	0.00	**0.58**
900	0.11	0.03	0.45	0.04	0.04	-0.11	0.00	0.00	**0.48**
950	0.09	0.02	0.90	0.04	0.04	-0.08	0.00	0.00	**0.91**
1000	0.08	0.01	0.75	0.04	0.04	-0.01	0.00	0.00	**0.76**
1050	0.08	0.01	0.87	0.04	0.04	-0.02	0.00	0.00	**0.87**
1100	0.06	0.01	0.78	0.04	0.04	-0.01	0.00	0.00	**0.78**
1150	0.06	0.01	0.96	0.04	0.04	-0.01	0.00	0.00	**0.96**
1200	0.05	0.01	1.13	0.04	0.04	-0.01	0.00	0.00	**1.13**
1250	0.04	0.01	1.30	0.04	0.04	-0.01	0.00	0.00	**1.30**
1300	0.04	0.01	1.28	0.04	0.04	-0.01	0.00	0.00	**1.28**
1350	0.04	0.01	1.73	0.04	0.04	0.00	0.00	0.00	**1.73**
1400	0.01	0.02	2.40	0.04	0.04	-0.01	0.00	0.00	**2.40**
1450	0.02	0.01	2.31	0.04	0.04	-0.01	0.00	0.00	**2.32**
1500	0.03	0.01	2.83	0.04	0.04	0.00	0.00	0.00	**2.83**
1550	0.26	0.02	2.46	0.04	0.04	0.01	0.00	0.00	**2.48**
1600	0.30	0.02	2.14	0.04	0.04	0.02	0.00	0.00	**2.17**
1650	0.35	0.03	2.52	0.04	0.04	0.09	0.00	0.00	**2.54**
1700	0.45	0.05	2.46	0.04	0.04	0.33	0.01	0.01	**2.52**
1750	0.52	0.13	2.61	0.04	0.04	0.30	0.07	0.00	**2.69**
1800	0.54	0.57	2.49	0.04	0.04	0.28	0.54	0.00	**2.68**

7.3.4 Filtered Detector Transfer

Special tests of filtered detectors (such as photometers) can be made. It should be noted that the relative measurement uncertainty (measurement statistical uncertainty) will increase due to the decrease in signal in the "wings" of the response. As mentioned above, when the slope of the responsivity curve is steep, the uncertainties due to stray light, bandwidth, and wavelength calibration can increase by orders of magnitude relative to the uncertainties within the bandpass of the filtered detector.

7.4 Spatial Uniformity Measurement Uncertainty

The repeatability uncertainty components for the responsivity uniformity measurement of a typical photodiode are listed in table 7.8. This is the repeatability of the measured relative responsivity in the central portion of the active area during one measurement scan. The measurement noise is the average standard deviation of the mean of the measurements in the central portion of the active area of the detector. The relative combined standard uncertainty is the RSS of the measurement noise and the one-day DVM uncertainty specification.

The measurement repeatability uncertainty depends on the SNR of the detector and noise due to the amplifier and DVM. The relative measurement noise varies spectrally, primarily due to the change in monochromator flux magnitude with wavelength. Note that the variation in responsivity over the measured area is much larger than this uncertainty value.

The reported uncertainty is not an indication of the uniformity measurement reproducibility. It does not consider other components which contribute to the reproducibility uncertainty, such as, the ability to reproduce the same irradiance geometry and detector alignment. Uniformity reproducibility results have been reported [49] for a Hamamatsu S1337-1010BQ with a standard deviation of 0.033 % at 500 nm and 0.25 % at 1000 nm.

The intended primary use of the reported uniformity results is qualitative. That is, to indicate if any large discontinuities are present in the responsivity uniformity which can lead to larger than expected uncertainties in absolute responsivity measurements. Quantitative application of the reported uniformity results requires examination of the irradiance geometry and equipment involved. The generalized application of the uniformity measurement results is currently being studied.

Table 7.8. Photodetector Spatial Uniformity Measurement Repeatablity Uncertainty

Source of uncertainty	Relative measurement noise[†]	DVM uncertainty	Relative combined standard uncertainty [%]
Type	A	B	
Relative uncertainty	$u(R)/R$	$u(V)/V$	$u_c(S_{Unif})/S_{Unif}$
	Estimated value [%]		Root-sum-of-squares
	0.0009	0.0007	0.0012

[†]Depends on photodetector and signal level.

8. Quality System

The spectroradiometric detector measurements described in this publication are part of the NIST Optical Technology Division calibration services and are in compliance with ANSI/NCSL Z540-1-1994 [57, 58, 59]. Although quality procedures were previously in place, they varied from calibration service to calibration service within the Division. The quality control procedures were typically limited to the technical aspects of the measurements, such as the yearly calibration of voltmeters, the use of multiple working standards, and their routine rotation.

The goal of the Optical Technology Division's ANSI/NCSL Z540-1-1994 compliance project (which began in 1993) was to unify all the calibration services offered by the Division with standard formats and similar procedures. Balancing functionality and bureaucracy was a concern from the start. Efforts were directed toward developing a useful and practical quality system. Excessively sophisticated and complex procedures are avoided, along with redundant documentation. Tools such as checklists, forms, and flowcharts are used where applicable.

8.1 Control Charts

Control charts are a standard statistical tool used for tracking a process over time [60]. Two working standard detectors are used for each test (customer) detector measurement. Thus, one working standard can be used to calculate the spectral responsivity of the second working standard. Since the working standards are randomly chosen each week, this allows all of the working standards to be compared to each other over time. The responsivity of a given working standard detector can then be tracked over time using control charts. In practice, only a few wavelengths (e.g., every 100 nm) need to be plotted on a control chart for each working standard. An example control chart for a visible working standard at 600 nm is shown in figure 8.1. The center line is the mean responsivity during a period when the measurement process is stable (i.e., "in control"). The upper control limit (UCL) and lower control limit (LCL) are respectively plus and minus three times the average standard deviation of the mean of the responsivity measurements. This should include almost all of the expected random measurement fluctuations. Figure 8.1 shows a trend in the responsivity measurements that does not appear to be due to random fluctuations but is still within the control limits. This demonstrates the value of control charts.

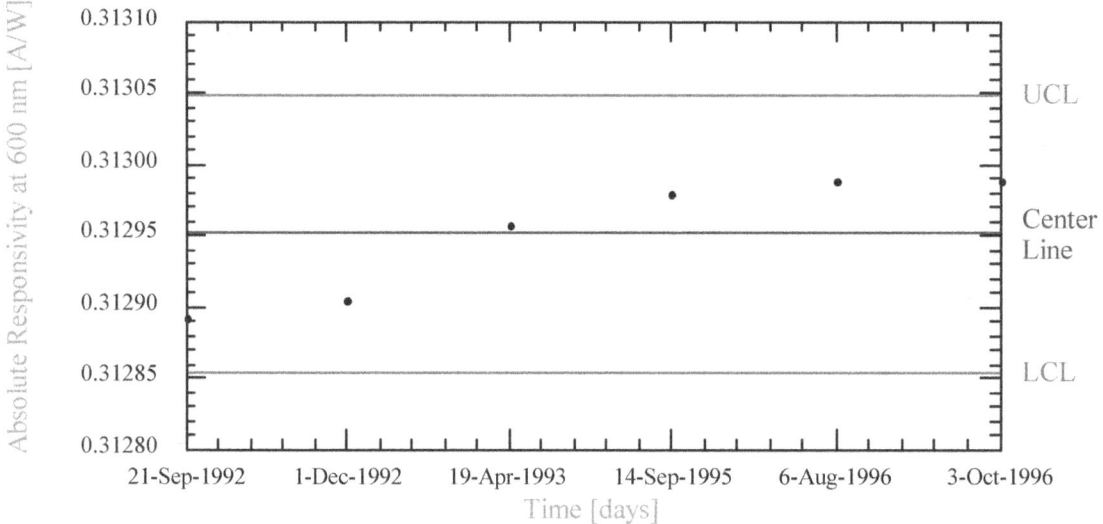

Figure 8.1. Control chart example for a NIST Visible Working Standard (Vis WS).

8.2 Comparison to Other Laboratories

Comparisons have been made between the UV responsivity scales of the UV SCF and the NIST Synchrotron Ultraviolet Radiation Facility (SURF) II [61, 62]. The SURF II responsivity scale is derived from a different physical basis (synchrotron radiation) than the UV SCF's scale (derived from the HACR). The comparisons have agreed within the stated uncertainties of both scales. Future UV intercomparisons are planned with SURF III [63] and a second absolute cryogenic radiometer being installed at SURF III.

There have been few international spectral responsivity intercomparisons. The latest carried out by the Bureau International des Poids et Measures (BIPM) was in 1992 [32, 64]. NIST agreed with the BIPM responsivity scale to better than 0.01 % from 500 nm to 900 nm. The relative difference increased to $\approx$ 0.4 % at 950 nm and 1000 nm and steadily increased from 0.03 % at 450 nm to 3 % at 248 nm. The larger deviations below 400 nm were reported by several labs (including NIST) but were typically within the stated measurement uncertainties.

Currently there is a spectral responsivity comparison underway between NIST and Centro Nacional De Metrología (CENAM) in Mexico. It is hoped that with the establishment of the North American Calibration Cooperation (NACC) this will be an ongoing intercomparison and will eventually include National Research Council (NRC) in Canada as well. It is also expected that there will be additional intercomparisons among regional groups of Europe, Asia, North America, and South America.

9. Characteristics of Photodiodes Available from NIST

This section describes the characteristics of the silicon photodiodes provided by NIST under Service ID numbers 39071S and 39073S. The physical and electrical characteristics of the photodiodes are discussed and the results of linearity measurements are given. The precision apertures are described and the photodiode fixture mechanical diagrams are shown.

9.1   Hamamatsu S1337-1010BQ

Hamamatsu S1337-1010BQ silicon photodiodes have been supplied by NIST as spectral (power) responsivity standards. NIST currently supplies the Hamamatsu S2281 for this purpose (NIST Service ID number 39073S). As mentioned previously, the Hamamatsu S1337 series diode is a popular diode for radiometric standards and it has been extensively characterized [31, 32]. Hamamatsu describes [65] the S1337-1010BQ as a p-n diode with a 1 cm x 1 cm active area, a fused quartz window, and a ceramic case. The spectral response range is 190 nm to 1100 nm with a peak at 960 nm. The S1337-1010BQ also has a high shunt resistance (dynamic impedance), with a typical value of 200 MΩ and a minimum value of 50 MΩ.

The typical measured spectral responsivity and quantum efficiency are shown in figures 6.2 and 6.3, respectively. The typical spatial uniformities measured at 500 nm and 1000 nm are shown in figure 6.4a and b, respectively. The temperature coefficient of several and S1337 series photodiodes were measured using a temperature-controlled fixture. All of the measurements were made following the typical spectral responsivity procedures at temperatures around 25 °C. Figure 9.1 shows the average temperature coefficient of the Hamamatsu S1337 series photodiode. A second common silicon photodiode series, the S1226, is shown for comparison in figure 9.1 along with the wavelength of peak responsivity for each photodiode.

The linearity of the S1337-1010BQ at 633 nm is shown in figure 9.2 spanning irradiance levels from 0.5 mW/cm^2 to 6.6 mW/cm^2. Each data point represents the ratio of the photodiode responsivity at the indicated irradiance to the responsivity at low power. The linearity was measured by using a beamsplitter to irradiate two photodiodes at approximately a 10:1 intensity ratio, with the diode aperture filled and uniformly irradiated [66]. The linearity is dependent on the irradiation geometry and will differ for spot sizes significantly smaller than the aperture size.

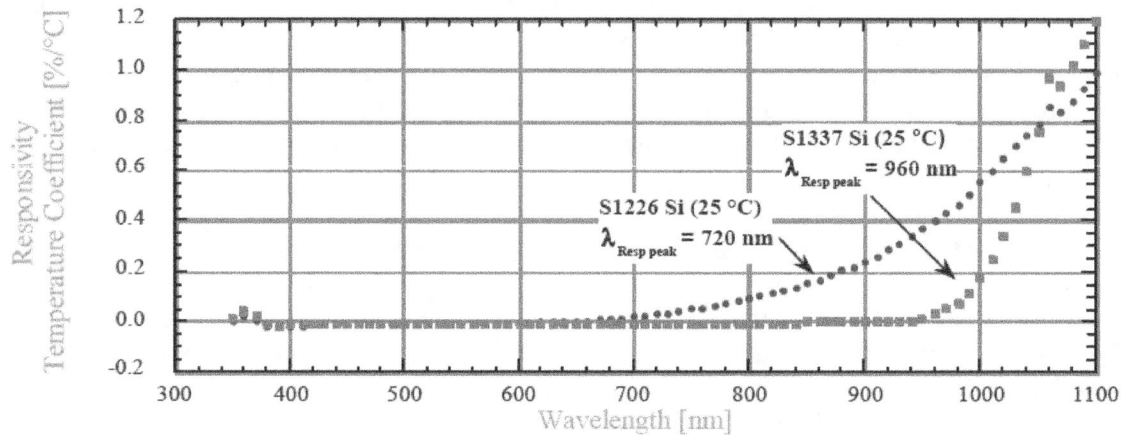

Figure 9.1. Temperature coefficient of silicon Hamamatsu S1226 and S1337 photodiodes.

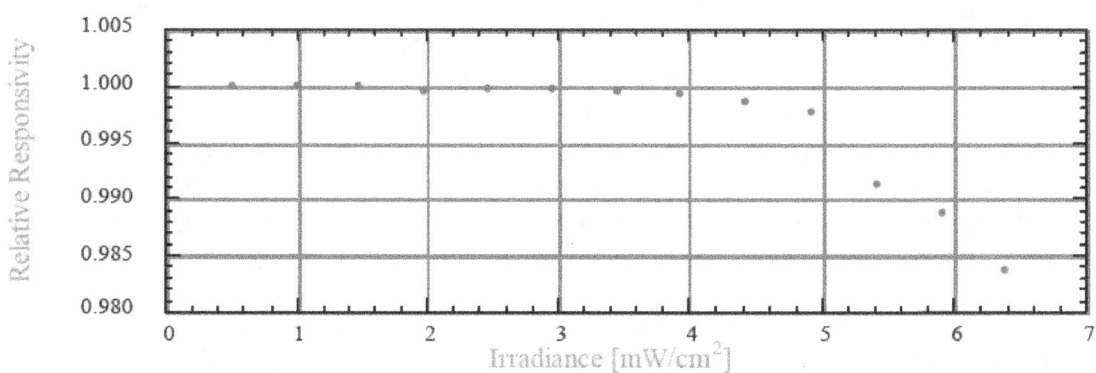

Figure 9.2. Linearity of Hamamatsu S1337-1010BQ at 633 nm.

9.2 Hamamatsu S2281

Hamamatsu S2281 silicon photodiodes are currently provided by NIST as standards of spectral responsivity (NIST Service ID number 39073S). The characteristics are essentially identical to the S1337-1010BQ. Since the S2281 is almost indistinguishable from the S1337, the spectral responsivity and quantum efficiency are not shown in figures 6.2 and 6.3. The significant differences between the two types of diodes are the S2281 has a 1 cm^2 active area that is circular instead of square and it is housed in a metal case with a BNC connector. The BNC case simplifies the photodiode fixture since no electrical wiring is required.

9.3 UDT Sensors UV100

UDT Sensors UV100 silicon photodiodes are provided as spectral responsivity standards in the UV (NIST Service ID number 39071S). UDT Sensors literature [67] describes the UV100 as an inverted channel diode with enhanced resistance to damage from UV radiation. The UV100 has a quartz window, a 1 cm^2 circular active area, and is housed in a metal case with a BNC connector similar to the Hamamatsu S2281. The spectral response range is from 200 nm to 1100 nm with a peak around 760 nm. The typical shunt resistance value is 10 MΩ.

The linearity of the UV100, shown in figure 9.3 at 442 nm, spans irradiance levels from 0.1 mW/cm^2 to 1.1 mW/cm^2 with and without a reverse bias voltage. Each data point represents the ratio of the photodiode responsivity at the indicated irradiance to the responsivity within the linear region. The linearity was measured by using a beamsplitter to irradiate two photodiodes at approximately a 10:1 intensity ratio, with the diode aperture filled and uniformly irradiated [66]. The linearity is dependent on the irradiation geometry and will differ for spot sizes significantly smaller than the aperture size.

The change in the responsivity as a function of bias voltage for this type of photodiode at 442 nm is shown in figure 9.4. For wavelengths shorter than 450 nm, a 1 V bias can be used to improve the linearity of the photodiode without significantly changing the spectral responsivity. There will however be some leakage current which will limit the minimum usable signal.

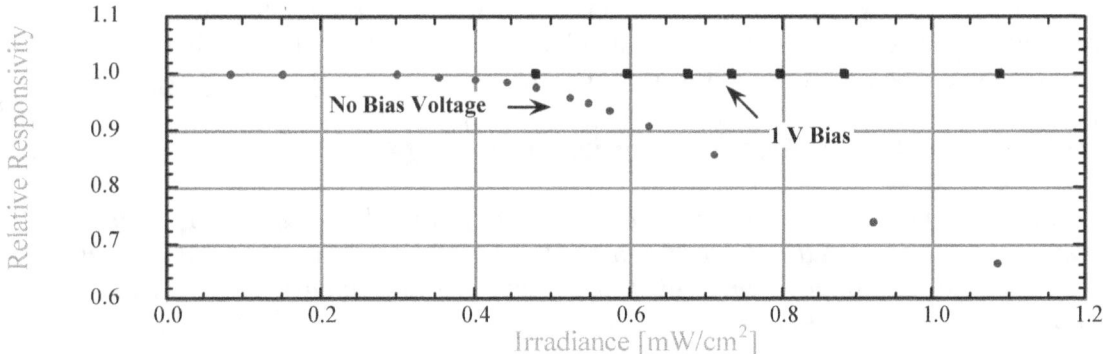

Figure 9.3. Linearity of UDT Sensors UV100 at 442 nm.

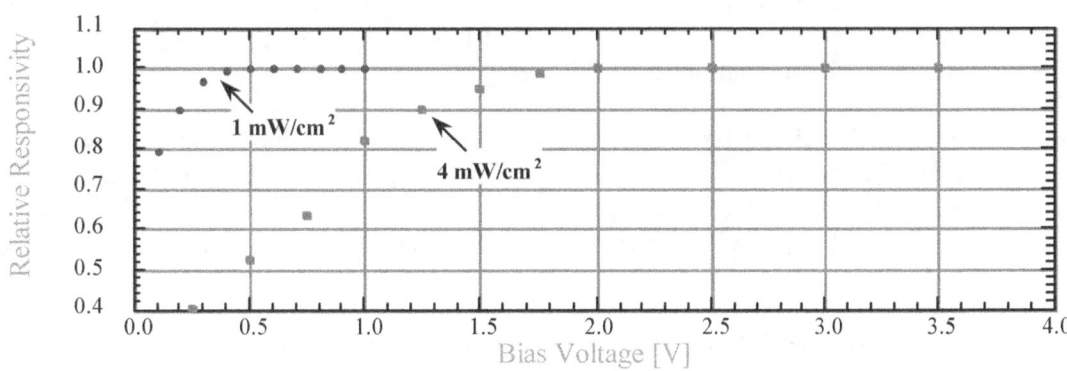

Figure 9.4. Responsivity dependence on bias voltage of UDT Sensors UV100 at 442 nm.

9.4 Detector Apertures

The precision apertures provided with the photodiodes by NIST Service ID numbers 39071S and 39073S are Buckbee Mears part number SK#030483-1073. They are electroformed optical apertures that have a bi-metal construction (copper with a nickel finish on both sides). The aperture is a thin disc with an outside diameter of 13.92 mm (0.5480 in) and an inside diameter of 7.9789 mm (0.31413 in), the specified aperture area is nominally 0.5 cm^2. The apertures are dimensionally measured at NIST in the Fabrication Technology Division in an environmentally controlled laboratory. The aperture diameter is typically measured twice along perpendicular axes. The reported aperture area is the average area calculated from the two (or more) diameters.

The reported aperture area and distance from the aperture plate to the photodiode surface do not enter into the responsivity measurement results because the optical beam underfills the aperture. The uncertainty values reported with the aperture dimensions are expanded uncertainties.

Recently an new method has been reported for measuring aperture areas [68]. This method determines the area by optical comparison and is faster than the mechanical diameter measurements. Some of the apertures provided have been measured in this manner. Soon, all of the apertures will be measured using this new method.

9.5 Detector Fixture Mechanical Drawings

Note: As stated in the Introduction, this document follows the NIST policy of using the International System of Units (SI). The following mechanical drawings were originally prepared in English units and are presented without converting the values shown to SI units.

The detector fixtures for the Hamamatsu S1337-1010BQ and S2281 and the UDT Sensors UV100 silicon photodiodes are designed for convenient handling and use. The fixture housings are black anodized aluminum and the 5.08 cm diameter was chosen as a convenient size for use with common optical table fixtures. Most of the fixtures also have a 1/4-20 threaded hole (not shown in drawings) on the side of the fixture for a standard optical table post. Each fixture has an engraved serial number on the back. Black anodized aluminum covers (not shown) were added later to protect the photodiodes (and apertures) during storage and shipment.

The Hamamatsu S1337-1010BQ fixture design includes space to couple the photodiode anode and cathode to a BNC connector. Figure 9.5 shows the mechanical diagram for the fixture body. Figure 9.6 shows additional pieces for holding the diode lead sockets and BNC connector. Figure 9.7 is an exploded view of the fixture including the photodiode, aperture, and aperture plate.

The UDT Sensors UV100 and Hamamatsu S2281 photodiodes are housed in metal BNC cases. This simplifies the fixture and reduces the construction time since no electrical wiring is required. The mechanical diagram for the UV100 and S2281 diodes is shown in figure 9.8.

The precision apertures are attached to a black anodized aluminum plate. Figure 9.9 shows the first design for the aperture plate. This was found to scatter light onto the diode when light struck the countersunk area around the aperture. An improved design shown in figure 9.10 has been used since 1993. Later productions of the aperture plate are engraved with the detector fixture serial number facing the diode.

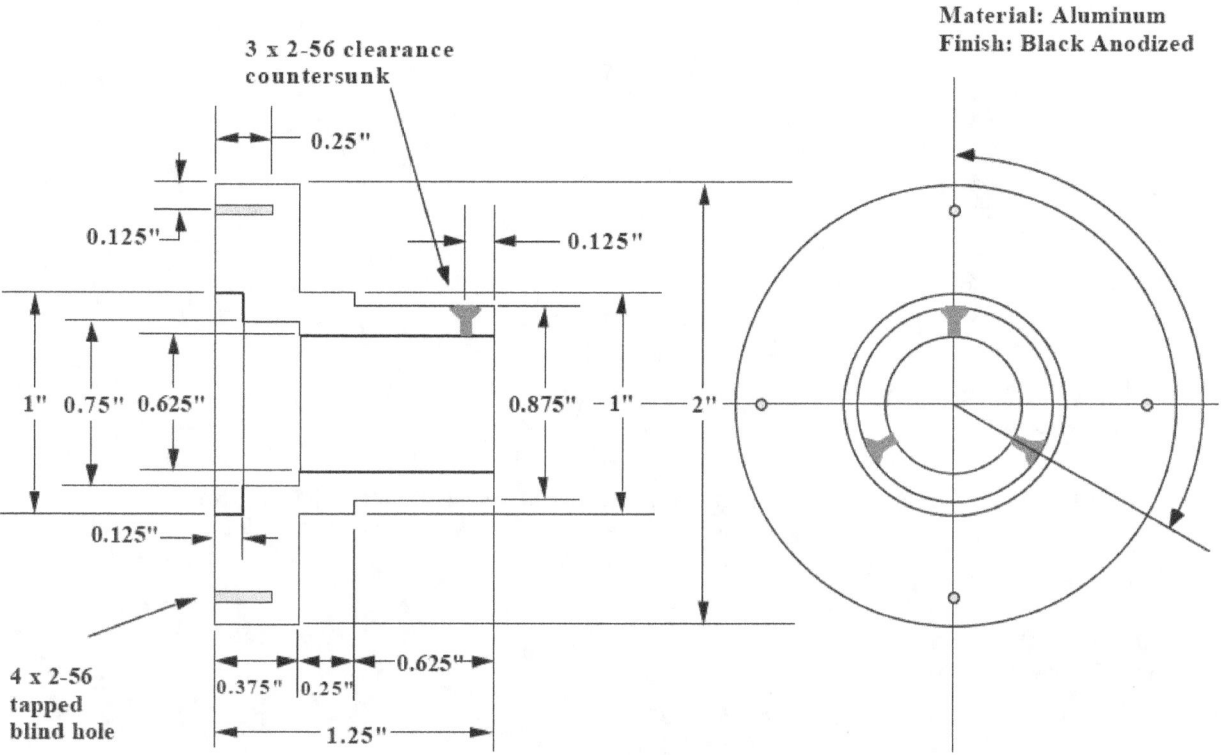

Figure 9.5. Mechanical diagram of Hamamatsu S1337-1010BQ fixture body.

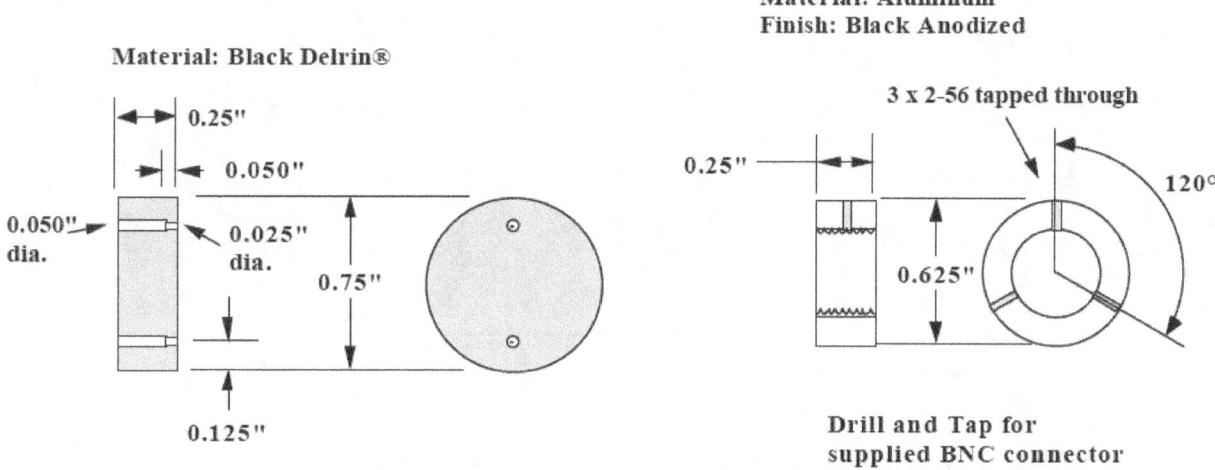

Figure 9.6. Mounting pieces for the Hamamatsu S1337-1010BQ photodiode and BNC connector.

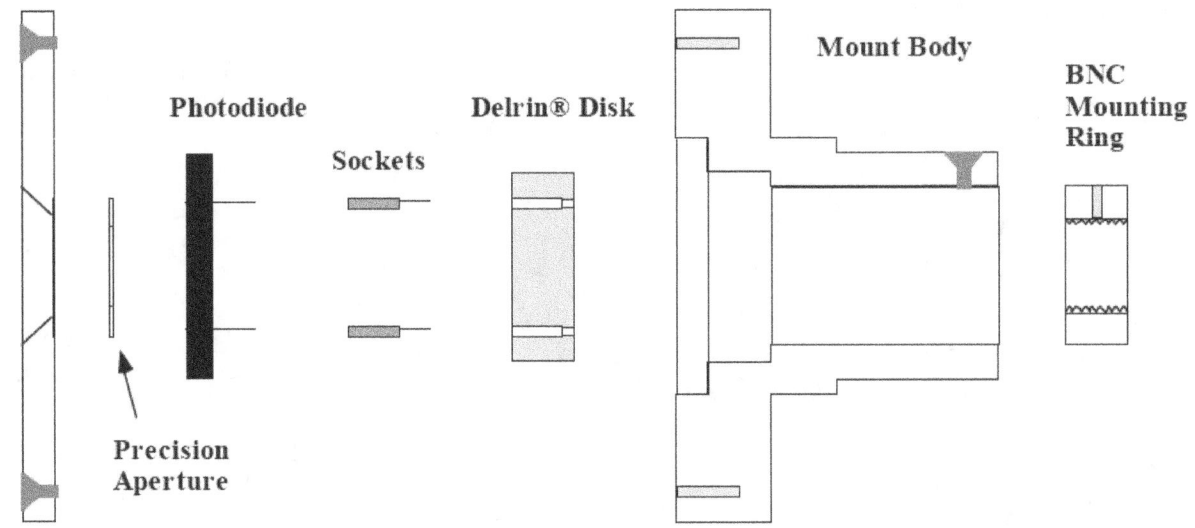

Figure 9.7. Exploded view diagram of Hamamatsu S1337-1010BQ fixture.

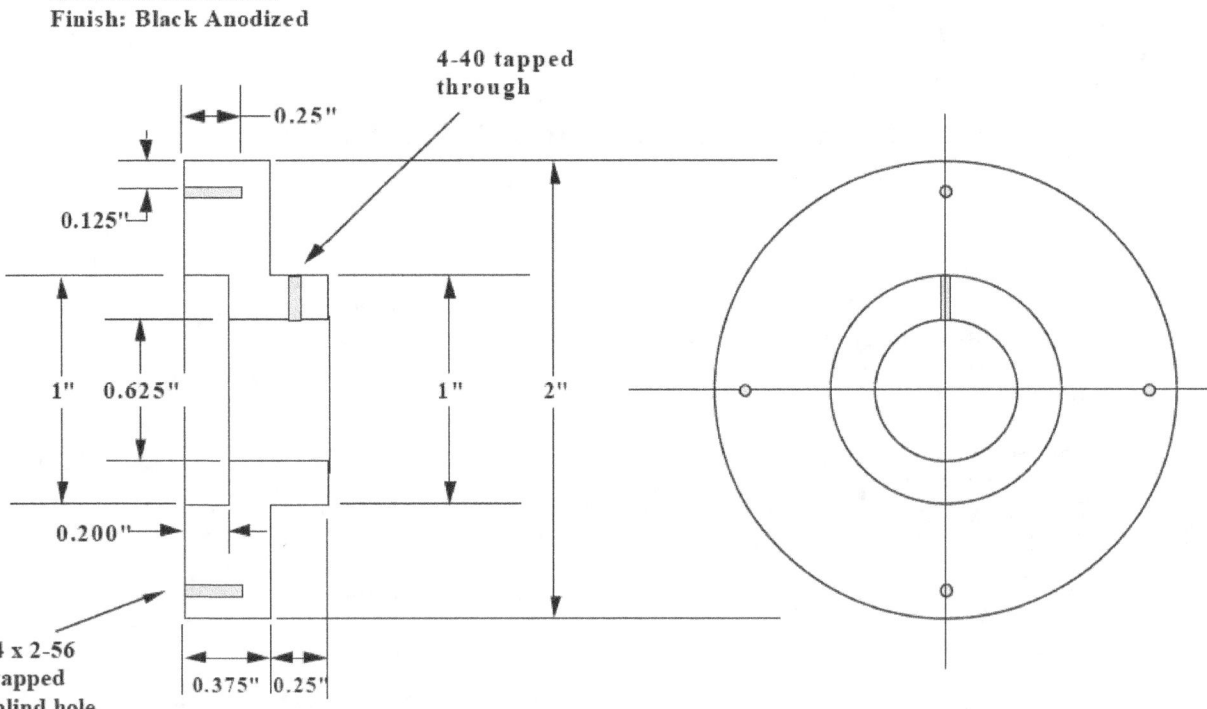

Figure 9.8. Mechanical diagram of UDT Sensors UV100 and Hamamatsu S2281 fixture.

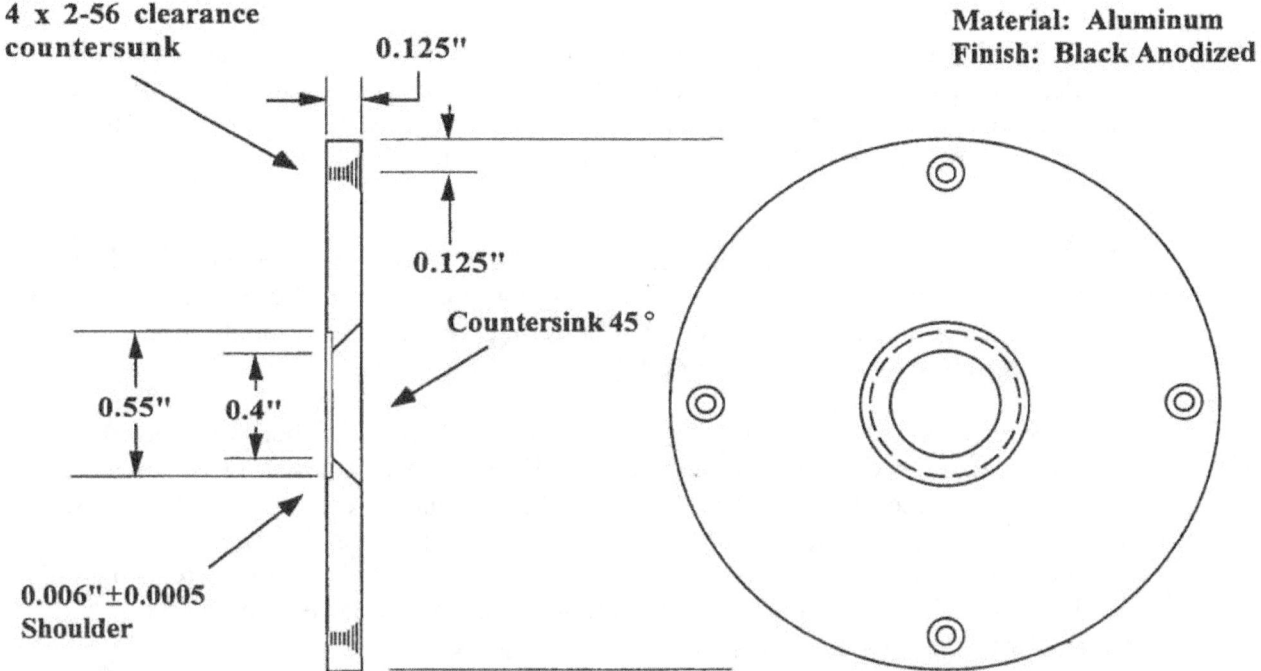

Figure 9.9. Mechanical diagram of pre-1993 aperture plate for detector fixtures.

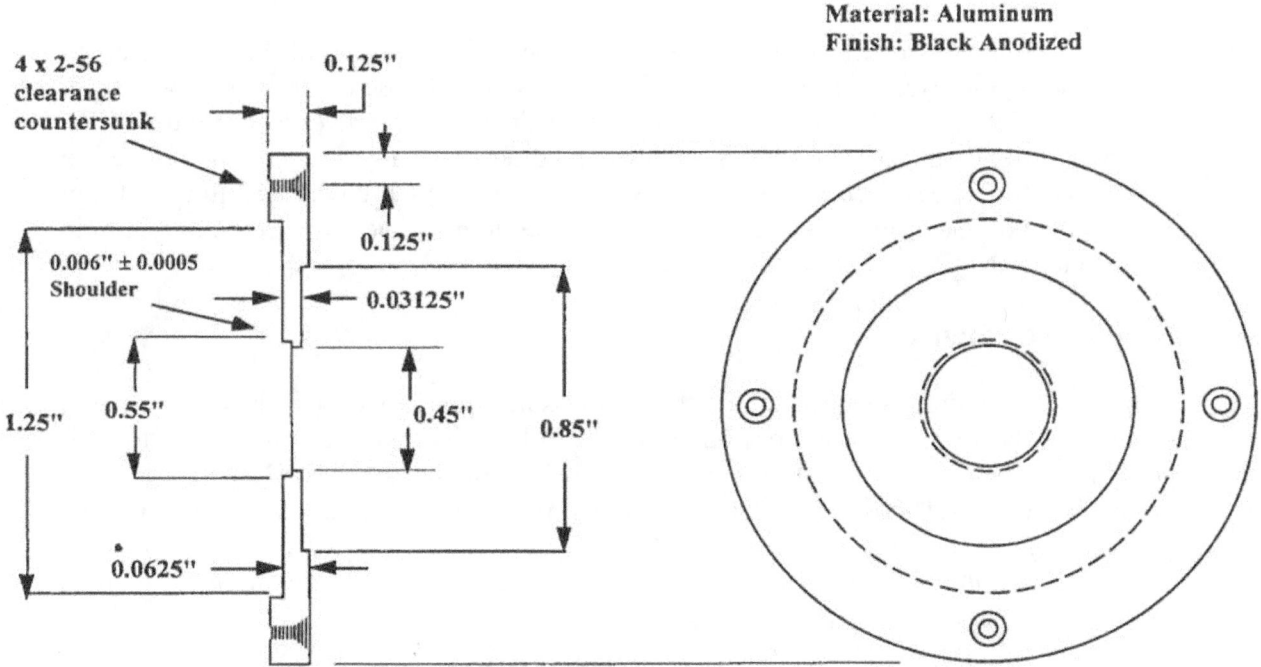

Figure 9.10. Mechanical diagram of present aperture plate for detector fixtures.

10. Future Work

Previous sections mentioned future modifications and improvements to the UV and Vis/NIR SCFs. This section will discuss these changes and other future work. Several publications are planned to describe these modifications and improvements to the SCFs. This publication describing the measurement services offered by the UV and Vis/NIR SCFs will be revised every 3 to 5 years.

One of the first additions will be to temperature control the UV and Vis/NIR SCFs' silicon working standards. A comparison of the UV SCF working standards to the SURF III [63] responsivity scale is planned in the near future. A long-term study is ongoing to find new UV detectors with better responsivity, uniformity, and stability.

Several major modifications will be made to the UV SCF. A new, higher efficiency UV monochromator (with automated order sorting filters) and larger linear translation stages have been purchased and will be integrated into the UV SCF when the measurement schedule allows. The translation stages will replace the rotating stage and small linear stages that have limited the UV SCF to measuring only one test detector at a time. The new stages will be configured similar to the Vis/NIR SCF stages. Additional imaging mirrors for astigmatism correction will be added to the UV SCF similar to the Vis/NIR SCF.

A higher accuracy, computer-addressable wavelength encoder will be added to the Vis/NIR SCF to improve the wavelength control. Additional modifications to improve the measurement service are high-accuracy aperture measurements [68], automated control of the amplifier gains, and a computer interface (via IEEE-488 bus) with customer-supplied test equipment.

As seen in the uncertainty assessment, the measurement uncertainty was significantly affected by the poor response of the pyroelectric detector in the SCFs. A flat detector with lower noise equivalent power (NEP) and higher SNR would be better suited for the power levels in the SCFs. There are efforts currently underway at NIST to develop better spectrally flat detectors. One is a silicon bolometer [69, 70] designed as a standard with the new IR SCF covering 2 µm to 20 µm [71, 72]. Since the bolometer has a flat response, it can also, in principle, be used below 2 µm in the UV, visible, and near-IR.

11. Acknowledgments

The development, operation, and evolution of this measurement service has involved several NIST personnel. The major contributors have been: Bob Saunders and Jeanne Houston for constructing the Vis/NIR SCF; Ed Zalewski and Jeanne Houston for their work establishing and operating the Detector Response Transfer and Intercomparison Program (DRTIP); and Joel Fowler for the design of the DRTIP electronics. Chris Cromer added the UV SCF, enhanced the computer automation, and started the sales of silicon photodiodes as part of the measurement service.

The authors would like to thank Tom Gentile, Jeanne Houston, Jonathan Hardis, Chris Cromer, and Chris Classon for their work with the HACR and the traps; Joel Fowler for providing the transimpedance amplifiers; and Bob Saunders for his informative discussions on spectroradiometry and equipment. Thanks are also given to the staffs of the NIST Calibration Program and NVLAP for their help with the measurement service. In particular thanks go to Janet Land of the Precision Engineering Division and Carroll Brickenkamp of the Measurement Services office for their help with quality systems and documentation.

12. References

[1] American National Standard for Calibration - Calibration Laboratories and Measuring and Test Equipment - General Requirements, ANSI/NCSL Z540-1-1994. (Currently available through the National Conference of Standards Laboratories, 1800 30th Street, Suite 305B, Boulder, CO 80301.)

[2] Determination of the Spectral Responsivity of Optical Radiation Detectors, Publ. 64 (Commission Internationale de L'Éclairage, Paris, 1984). (Currently available through the U. S. National Committee of the CIE, c/o Mr. Thomas M. Lemons, TLA-Lighting Consultants, Inc., 72 Loring Avenue, Salem, MA 01970.)

[3] W. Budde, *Optical Radiation Measurements, Volume 4: Physical Detectors of Optical Radiation*, Academic Press, Inc., Orlando, FL (1983) pp. 38-59.

[4] J. Geist, L. B. Schmidt, and W. E. Case, "Comparison of the Laser Power and Total Irradiance Scales Maintained by the National Bureau of Standards," *Appl. Opt.* **12**, 2773-2776 (1973).

[5] J. Geist, M. A. Lind, A. R. Schaefer, and E. F. Zalewski, "Spectral Radiometry: A New Approach Based on Electro-Optics," Natl. Bur. Stand. (U.S.), Tech. Note 954 (1977).

[6] Unpublished preprint of sequel to NBS Technical Note 950 describing the radiometric characteristics of the DRTIP radiometers, p. 9.

[7] B. N. Taylor and C. E. Kuyatt, Guidelines for Evaluating and Expressing the Uncertainty of NIST Measurement Results, Natl. Inst. Stand. Technol., Tech. Note 1297 (1994 ed.).

[8] E. F. Zalewski and J. Geist, "Silicon Photodiode Absolute Spectral Response Self-Calibration," *Appl. Opt.* **19**, 1214-1216 (1980).

[9] J. Geist, E. F. Zalewski, and A. R. Schaefer, "Spectral Response Self-Calibration And Interpolation of Silicon Photodiodes," *Appl. Opt.* **19**, 3795-3799 (1980).

[10] E. F. Zalewski and C. R. Duda, "Silicon Photodiode Device with 100 % External Quantum Efficiency," *Appl. Opt.* **22**, 2867-2873 (1983).

[11] The QED-200 is now manufactured by UDT Instruments (formerly Graseby Optronics), Orlando, FL.

[12] E. F. Zalewski, The NBS Photodetector Spectral Response Calibration Transfer Program, Natl. Bur. Stand. (U.S.), Spec. Publ. 250-17, p. 45 (1988).

[13] J. M. Houston and E. F. Zalewski, "Photodetector Spectral Response Based on 100 % Quantum Efficient Detectors," *Optical Radiation Measurements II*, James M. Palmer, Editor, Proc. SPIE 1109, pp. 268-277 (1989).

[14] C. L. Cromer, "A New Spectral Response Calibration Method using a Silicon Photodiode Trap Detector," presented at the 1991 Measurement Science Conference, Anaheim, CA, Jan. 31-Feb. 1, 1991 (unpublished).

[15] N. P. Fox, "Trap Detectors and Their Properties," *Metrologia* **28**, 197-202 (1991).

[16] J. Geist, "Current Status of, and Future Directions In, Silicon Photodiode Self-Calibration," *Optical Radiation Measurements II*, James M. Palmer, Editor, Proc. SPIE 1109, 246-256 (1989).

[17] E. F. Zalewski and C. C. Hoyt, "Comparison Between Cryogenic Radiometry and the Predicted Quantum Efficiency of pn Silicon Photodiode Light Traps," *Metrologia* **28**, 203-206 (1991).

[18] N. P. Fox, "Radiometry with Cryogenic Radiometers and Semiconductor Photodiodes," *Metrologia* **32**, 535-543 (1995/96).

[19] A. C. Parr, A National Measurement System for Radiometry, Photometry, and Pyrometry Based Upon Absolute Detectors, Natl. Inst. Stand. Technol., Tech. Note 1421 (1996).

[20] T. R. Gentile, J. M. Houston, J. E. Hardis, C. L. Cromer, and A. C. Parr, "National Institute of Standards and Technology High-Accuracy Cryogenic Radiometer," *Appl. Opt.* **35**, 1056-1068 (1996).

[21] T. R. Gentile, J. M. Houston, and C. L. Cromer, "Realization of a Scale of Absolute Spectral Response Using the National Institute of Standards and Technology High-Accuracy Cryogenic Radiometer," *Appl. Opt.* **35**, 4392-4403 (1996).

[22] T. C. Larason, S. S. Bruce, and C. L. Cromer, "The NIST High Accuracy Scale for Absolute Spectral Response from 406 nm to 920 nm," *J. Res. Natl. Inst. Stand. Technol.* **101**, 133-140, (1996).

[23] NIST Calibration Services Users Guide, Joe D. Simmons, Editor, Natl. Inst. Stand. Technol., Spec. Publ. 250 (1991).

[24] NIST Calibration Services Users Guide Fee Schedule, Natl. Inst. Stand. Technol., Spec. Publ. 250 Appendix (1997).

[25] H. J. Kostkowski and F. E. Nicodemus, An Introduction to the Measurement Equation, Chapter 5 in Self-Study Manual on Optical Radiation Measurements: Part I--Concepts, Chapters 4 and 5, F. E. Nicodemus, Editor, Natl. Bur. Stand. (U.S.), Tech. Note 910-2 (1978).

[26] H. J. Kostkowski, The Relative Spectral Responsivity and Slit-Scattering Function of a Spectroradiometer, Chapter 7 in Self-Study Manual on Optical Radiation Measurements: Part I--Concepts, Chapters 7, 8, and 9, F. E. Nicodemus, Editor, Natl. Bur. Stand. (U.S.), Tech. Note 910-4 (1979).

[27] R. Köhler, R. Goebel, R. Pello, and J. Bonhoure, "Effects of Humidity and Cleaning on the Sensitivity of Si Photodiodes," *Metrologia* **28**, 211-215 (1991).

[28] A description of diffraction and the Airy disk can be found in most basic optics texts, e.g., F. L. Pedrotti and L. S. Pedrotti, *Introduction to Optics*, 2nd edition, Prentice Hall, Englewood Cliffs, NJ (1993) or E. Hecht, *Optics*, 2nd edition, Addison-Wesley Publishing Company, Reading, MA (1987).

[29] J. B. Shumaker, Introduction to Coherence in Radiometry, Chapter 10 in Self-Study Manual on Optical Radiation Measurements: Part I--Concepts, Chapter 10, F. E. Nicodemus, Editor, Natl. Bur. Stand. (U.S.), Tech. Note 910-6 (1983).

[30] J. L. Gardner, "Astigmatism Cancellation in Spectroradiometry," *Metrologia* **28**, 251-254 (1991).

[31] F. Lei and J. Fischer, "Characterization of Photodiodes in the UV and Visible Spectral Region Based on Cryogenic Radiometry," *Metrologia* **30**, 297-303 (1993).

[32] R. Köhler, R. Goebel, and R. Pello, Report on the International Comparison of Spectral Responsivity of Silicon Detectors, Rapport BIPM-94/9, document CCPR/94-2, dated July 27, 1994, Bureau International des Poids et Mesures, Pavillon de Breteuil, 93212 Sevres, Cedex, France.

[33] J. M. Bridges and W. R. Ott, "Vacuum Ultraviolet Radiometry. 3: The Argon Mini-Arc as a New Secondary Standard of Spectral Radiance," *Appl. Opt.* **16**, 367-376 (1977).

[34] G. Eppeldauer and J. E. Hardis, "Fourteen-Decade Photocurrent Measurements with Large-Area Silicon Photodiodes at Room Temperature," *Appl. Opt.* **30**, 3091-3099 (1991).

[35] W. Budde, *Optical Radiation Measurements, Volume 4: Physical Detectors of Optical Radiation*, Academic Press, Inc., Orlando, FL (1983) pp. 265-268.

[36] G. Eppeldauer, "Electronic Characteristics of Ge and InGaAs Radiometers," *Infrared Technology and Applications XXII*, Björn F. Andresen and Marija Strojnik, Editors, Proc. SPIE 3061, pp. 833-838 (1997).

[37] J. E. Martin, N. P. Fox, and P. J. Key, "A Cryogenic Radiometer for Absolute Radiometric Measurements," *Metrologia* **21**, 147-155 (1985).

[38] At the 1994 CCPR meeting in Paris approximately 10 nations indicated that they had purchased a cryogenic radiometer or were in the process of purchasing one and would use it as the basis of the radiometric measurements in their respective nations.

[39] Electrical standards are maintained in the Electronics and Electrical Engineering Laboratory at NIST.

[40] R. C. Paule and J. Mandel, "Consensus Values and Weighting Factors," *J. Res. Natl. Bur. Stand.* (U.S.) **87**, 5 (1982).

[41] There are several books on optical detectors and radiometry that describe pyroelectric detectors. Many of these books are listed in the bibliography along with the following.

E. L. Dereniak and D. G. Crowe, *Optical Radiation Detectors*, Wiley, New York, NY (1984).

W. Budde, *Optical Radiation Measurements, Volume 4: Physical Detectors of Optical Radiation*, Academic Press, Inc., Orlando, FL (1983).

[42] V. R. Weidner and J. J. Hsia, NIST Measurement Services: Spectral Reflectance, Natl. Bur. Stand. (U.S.), Spec. Publ. 250-8 (1987).

P. Y. Barnes and E. A. Early, NIST Measurement Services: Spectral Reflectance, Natl. Inst. Stand. Technol., Spec. Publ. 250-8 Revised (in preparation).

[43] K. L. Eckerle, J. J. Hsia, K. D. Mielenz, and V. R. Weidner, NIST Measurement Services: Regular Spectral Transmittance, Natl. Bur. Stand. (U.S.), Spec. Publ. 250-6 (1987).

[44] See Ref. 27 and personal communication with one of the authors. Note: Cleaning a photodiode can change its responsivity.

[45] E. M. Gullikson, R. Korde, L. R. Canfield, R. E. Vest, "Stable Silicon Photodiodes for Absolute Intensity Measurements in the VUV and Soft X-Ray Regions," *J. Elect. Spect. Rel. Phenom.* **80**, 313-316 (1996).

[46] K. D. Stock, "Internal Quantum Efficiency of Ge Photodiodes," *Appl. Opt.* **27**, 12-14 (1988).

[47] A. R. Schaefer, E. F. Zalewski, and J. Geist, "Silicon Detector Nonlinearity and Related Effects," *Appl. Opt.* **22**, 1232-1236 (1983).

[48] Private communication with Richard Austin of Gamma Scientific, San Diego, CA.

[49] T. C. Larason and S. S. Bruce, "Spatial Uniformity of Responsivity for Silicon, Gallium nitride, Germanium, and Indium Gallium Arsenide Photodiodes," *Metrologia* (to be published).

[50] E. F. Zalewski, Radiometry and Photometry, Chapter 24 in *Handbook of Optics*, Vol. II, 2nd edition, M. Bass, Editor-in-Chief, McGraw-Hill, New York, NY (1995).

[51] C. L. Wyatt, *Radiometric Calibration: Theory and Methods*, Academic Press, San Diego, CA (1978).

[52] American National Standard for Expressing Uncertainty - US Guide to the Expression of Uncertainty in Measurement, ANSI/NCSL Z540-2-1997. (Currently available through the National Conference of Standards Laboratories, 1800 30th Street, Suite 305B, Boulder, CO 80301.)

[53] A. Corrons and E. F. Zalewski, Detector Spectral Response from 350 to 1200 nm Using a Monochromator Based Spectral Comparator, Natl. Bur. Stand. (U.S.), Tech. Note 988 (1978).

[54] R. D. Saunders and J. B. Shumaker, "Apparatus Function of a Prism-Grating Double Monochromator," *Appl. Opt.* **25**, 3710-3714 (1986).

[55] N. M. Durant and N. P. Fox, "Evaluation of Solid-State Detectors for Ultraviolet Radiometric Applications," *Metrologia* **32**, 505-508 (1995/96).

R. Goebel, R. Köhler, and R. Pello, "Some Effects of Low-Power Ultraviolet Radiation on Silicon Photodiodes," *Metrologia* **32**, 515-518 (1995/96).

G. Eppeldauer, "Longterm Changes of Silicon Photodiodes and Their Use for Photometric Standardization," *Appl. Opt.* **29**, 2289-2294 (1990).

L. R. Canfield, J. Kerner, and R. Korde, "Stability and Quantum Efficiency Performance of Silicon Photodiode Detectors in the Far Ultraviolet," *Appl. Opt.* **28**, 3940-3943 (1989).

R. Korde and J. Geist, "Quantum Efficiency Stability of Silicon Photodiodes," *Appl. Opt.* **26**, 5284-5290 (1987).

K. D. Stock, "Spectral Aging Pattern of Carefully Handled Silicon Photodiodes," *Measurement* **5**, 141-144 (1987).

K. D. Stock and R. Heine, "On the Aging of Photovoltaic Cells," *Optik* (Weimar) **71**, 137-142 (1985).

[56] A. L. Migdall and C. Winnewisser, "Linearity of a Silicon Photodiode at 30 MHz And Its Effect on Heterodyne Measurements," *J. Res. Natl. Inst. Stand. Technol.* **96**, 143-146 (1991).

J. L. Gardner and F. J. Wilkinson, "Response Time and Linearity of Inversion Layer Silicon Photodiodes," *Appl. Opt.* **24**, 1531-1534 (1985).

A. R. Schaefer, E. F. Zalewski, and J. Geist, "Silicon Detector Nonlinearity and Related Effects," *Appl. Opt.* **22**, 1232-1236 (1983).

W. Budde, "Multidecade Linearity Measurements on Si Photodiodes," *Appl. Opt.* **18**, 1555-1558 (1979).

[57] T. C. Larason, "The Radiometric Physics Division's Efforts at Building a Quality System Based on ISO/IEC Guide 25," presented at the Asociacion Mexicana De Metrologia, A. C. 1994 Conference, Acapulco, Mexico May 10-13, 1994 (unpublished).

[58] S. S. Bruce and T. C. Larason, "Building a Quality System Based on ANSI/NCSL Z540-1-1994 - An Effort by the Radiometric Physics Division at NIST," *Proc. NCSL 1995 Workshop and Symposium* (National Conference of Standards Laboratories), Dallas, TX July 16-20, 1995.

[59] S. S. Bruce and T. C. Larason, Developing quality system documentation based on ANSI/NCSL Z540-1-1994 -- the optical technology division's effort, Natl. Inst. Stand. Technol., Internal Report 5866 (1996).

[60] Control charts are discussed in many texts on quality control and statistics, e.g., W. Mendenhall and T. Sincich, *Statistics for Engineering and the Sciences*, 3rd edition, Dellen Publishing Company, San Francisco, CA (1992).

[61] L. R. Canfield, "New Far UV Detector Calibration Facility at the National Bureau of Standards," *Appl. Opt.* **26**, 3831-3837 (1987).

[62] L. R. Canfield and N. Swanson, Far Ultraviolet Detector Standards, Natl. Bur. Stand. (U.S.), Spec. Publ. 250-2 (1987).

[63] NIST is currently upgrading SURF II, creating SURF III. Improvements will be made in many areas, including beam current monitoring, magnetic structure uniformity, and electron energy.

[64] R. Köhler, R. Goebel, and R. Pello, "Results of an International Comparison of Spectral Responsivity of Silicon Photodetectors," *Metrologia* **32**, 463-468 (1995/96).

[65]　These values are from the manufacturer's catalog, Photodiodes, Cat. No. KPD 0001E05, Aug. 1996 T, Hamamatsu Photonics K. K., Solid State Division, 1126-1, Ichino-cho, Hamamatsu City, 435-91, Japan

[66]　R. D. Saunders and J. B. Shumaker, "Automated Radiometric Linearity Tester," *Appl. Opt.* **23**, 3504-3506 (1984).

[67]　These values are from the manufacturer's catalog (obtained Å 1993), Optoelectronic Components Catalog, UDT Sensors, Inc., 12525 Chadron Ave., Hawthorne, CA, USA 90250

[68]　J. B. Fowler and G. Dezsi, "High Accuracy Measurement of Aperture Area Relative to a Standard Known Aperture," *J. Res. Natl. Inst. Stand. Technol.* **100**, 277-283 (1995).

[69]　G. Eppeldauer, A. L. Migdall, and C. L. Cromer, "Characterization of a High Sensitivity Composite Silicon Bolometer," *Metrologia* **30**, 317-320 (1993).

[70]　G. Eppeldauer, A. L. Migdall, and C. L. Cromer, "A Cryogenic Silicon Resistance Bolometer for Use as an Infrared Transfer Standard Detector," *Thermal Phenomena at Molecular and Microscales and in Cryogenic Infrared Detectors*, edited by M. Kaviany et al., (ASME HTD-Vol. 277, New York, NY, 1994), pp. 63-67.

[71]　G. Eppeldauer, "Near Infrared Radiometer Standards," *Optical Radiation Measurements III*, James M. Palmer, Editor, Proc. SPIE 2815, pp. 42-54 (1996).

[72]　A. L. Migdall and G. Eppeldauer, NIST Measurement Services: Spectroradiometric Detector Measurements: Part III -- Infrared Detectors, Natl. Inst. Stand. Technol., Spec. Publ. 250-42 (in preparation).

13. Bibliography

Optics

E. Hecht, *Optics*, 2nd edition, Addison-Wesley Publishing Company, 1987.

F. L. Pedrotti and L. S. Pedrotti, *Introduction to Optics*, 2nd edition, Prentice Hall, Englewood Cliffs, NJ, 1993.

F. A. Jenkins and H. E. White, *Fundamentals of Optics*, 4th edition, McGraw-Hill, New York, NY, 1976.

M. Born and E. Wolf, *Principles of Optics*, 6th (corrected) edition, Pergamon Press, Oxford, England, 1993.

Applied Optics and Optical Engineering, Vol. V, Optical Instruments, R. Kingslake, editor, Academic Press, San Diego, CA, 1969.

D. C. O'Shea, *Elements of Modern Optical Design*, Wiley, New York, NY, 1985.

W. J. Smith, *Modern Optical Engineering*, 2nd edition, McGraw-Hill, , New York, NY, 1990.

Military Handbook 141, Optical Design, U. S. Department of Defense, Washington, D. C., 1962.

Handbook of Optics, Vol. I and II, 2nd edition, M. Bass, editor-in-chief, McGraw-Hill, New York, NY, 1995.

The Optics Problem Solver, Research and Education Association, Piscataway, NJ, 1990 revision.

E. Hecht, *Schaum's Outline of Theory and Problems of Optics*, McGraw-Hill, New York, NY, 1975.

Detectors and Amplifiers

Determination of the Spectral Responsivity of Optical Radiation Detectors, CIE Publication 64 (Commission Internationale de l'Eclairage, Paris, 1984). Currently available through the U.S. National Committee of the CIE, c/o T. M. Lemons, TLA-Lighting Consultants, Inc., 72 Loring Ave., Salem, MA 01970.

E. L. Dereniak and D. G. Crowe, *Optical Radiation Detectors*, Wiley, New York, NY, 1984.

E. L. Dereniak and G. D. Boreman, *Infrared Detectors and Systems*, Wiley, New York, NY, 1996.

G. H. Rieke, *Detection of Light: from the Ultraviolet to the Submillimeter*, Cambridge University Press, New York, NY, 1994.

J. D. Vincent, *Fundamentals of Infrared Detector Operation and Testing*, Wiley, New York, NY, 1990.

T. M. Frederiksen, *Intuitive Operational Amplifiers*, McGraw-Hill, New York, NY, 1988.

J. G. Graeme, *Photodiode Amplifiers: Op Amp Solutions*, McGraw-Hill, New York, NY, 1996.

D. L. Terrell, *Op Amps: Design, Application, and Troubleshooting*, Butterworth-Heinemann, Newton, MA, 1996.

P. Horowitz and W. Hill, *The Art of Electronics*, 2nd edition, Cambridge University Press, New York, NY, 1989.

S. M. Sze, *Physics of Semiconductor Devices*, 2nd edition, Wiley, New York, NY, 1981.

S. M. Sze, *Semiconductor Devices, Physics and Technology*, Wiley, New York, NY, 1985.

Radiometry

Optical Radiation Measurements, F. Grum and C. J. Bartleson, editors, *Vol. 1: Radiometry*, F. Grum and R. J. Becherer; *Vol. 4: Physical Detectors of Optical Radiation*, W. Budde. Academic Press, San Diego, CA, Vol. 1: Vol. 4: 1983.

Absolute Radiometry: Electrically Calibrated Thermal Detectors of Optical Radiation, F. Hengstberger, editor, Academic Press, San Diego, CA, 1989.

C. L. Wyatt, *Radiometric Calibration: Theory and Methods*, Academic Press, San Diego, CA, 1978.

C. L. Wyatt, *Electro-Optical System Design: For Information Processing*, McGraw-Hill, New York, NY, 1991. Expanded and revised version of the author's *Radiometric System Design*.

W. R. McCluney, *Introduction to Radiometry and Photometry*, Artech House, Norwood, MA, 1994.

R. W. Boyd, *Radiometry and the Detection of Radiation*, Wiley, New York, NY, 1983.

R. H. Kingston, *Optical Sources, Detectors, and Systems: Fundamentals and Applications*, Academic Press, San Diego, CA, 1995.

R. H. Kingston, *Detection of Optical and Infrared Radiation*, Springer-Verlag, New York, NY, 1978.

New Developments and Applications in Optical Radiometry III, *Metrologia* **28**, (1991).

New Developments and Applications in Optical Radiometry IV, *Metrologia* **30**, (1993).

New Developments and Applications in Optical Radiometry V, *Metrologia* **32**, (1996).

Statistics/Error Analysis/Uncertainties

P. R. Bevington, *Data Reduction and Error Analysis for the Physical Sciences*, McGraw-Hill, New York, NY, 1969.

J. R. Taylor, *An Introduction to Error Analysis: The Study of Uncertainties in Physical Measurements*, University Science Books, Sausalito, CA, 1982.

ANSI/NCSL Z540-2-1997, U.S. Guide to the Expression of Uncertainty in Measurement. Currently available through the National Conference of Standard Laboratories (NCSL) Secretariat, 1800 30th Street, Suite 305B, Boulder, CO 80301.

B. N. Taylor and C. E. Kuyatt, Guidelines for Evaluating and Expressing the Uncertainty of NIST Measurement Results, Natl. Inst. Stand. Technol. (US), Tech. Note 1297 (1994 ed.).

W. Mendenhall and T. Sincich, *Statistics for Engineering and the Sciences*, 3rd edition, Dellen Publishing Company, San Francisco, CA, 1992.

NBS Handbook 91, Experimental Statistics, Natl. Bur. Stand. (U.S.), NBS HDBK 91, 1963.

Precision Measurement and Calibration, Harry Ku, editor, Natl. Bur. Stand. (U.S.), Spec. Publ. 300, Vol. 1, 1969.

Quality and Laboratory Accreditation

NIST Handbook 150, National Voluntary Laboratory Accreditation Program Procedures and General Requirements, James L. Cigler and Vanda R. White, Editors, 1994.

NIST Handbook 150-2, NVLAP Calibration Laboratories Technical Guide, 1994.

ANSI/NCSL Z540-1-1994, Calibration Laboratories and Measuring and Test Equipment - General Requirements. Currently available through the National Conference of Standard Laboratories (NCSL) Secretariat, 1800 30th Street, Suite 305B, Boulder, CO 80301.

International Organization for Standardization (ISO) / International Electrotechnical Commission (IEC), General Requirements for the Competence of Calibration and Testing Laboratories Guide 25 (1990). Currently available through the American National Standards Institute (ANSI), 11 West 42nd Street, 13th Floor, New York, NY 10036.

REPORT OF TEST

NIST Test # 39071S - Spectral Responsivity

for

UDT Sensors UV100 Silicon Photodiode, U1xxx

Submitted by:

Any Company
Mr. Daniel Doe
123 Calibration Court
Measurement City, MD 00000-0000

(See your Purchase Order No. XXXX-XX, dated January 1, 1997)

1. Description of Test Material

The test photodiode, labeled U1xxx, is a UDT Sensors UV100 inverted layer silicon photodiode in an anodized aluminum mount with a removable precision aperture and a BNC connector. The active area of the photodiode is ≈ 1 cm^2.

2. Description of Test

The test photodiode was compared to two silicon photodiode working standards, U5xx and U5xx, using the NIST Ultraviolet (UV) monochromator-based comparator facility [1] from 200 nm to 500 nm in 5 nm increments. The spectral comparisons between the test photodiode and working standard photodiodes were performed using a double monochromator and an argon arc as the tunable monochromatic source.

The circular exit aperture of the UV monochromator was imaged ($\approx f/5$) on the test photodiode, resulting in a beam diameter at the photodiode of 1.5 mm. The beam was centered on, and underfilled, the aperture.

The wavelength scale of the monochromator was calibrated with several laser and emission lines and is accurate to $\pm$ 0.1 nm over the entire spectral range. The bandpass of the monochromator was 4 nm. The short-circuit photocurrent from the test photodiode and each working standard photodiode was measured with a calibrated transimpedance amplifier. The test photodiode and each working standard photodiode were measured with zero bias voltage. Beam power fluctuations were monitored with a beamsplitter and silicon photodiode. The absolute spectral responsivity scale is based on a high accuracy cryogenic radiometer, with a relative expanded uncertainty ($k = 2$) to absolute (SI) units of 0.2 %.

The spatial uniformity of the responsivity across the test photodiode photosensitive area was measured at 350 nm using the described comparator facility. The uniformity was measured in 0.5 mm increments using a 1.5 mm diameter beam.

Laboratory Environment:
   Temperature: 23.x °C $\pm$ 0.8 °C

Test Date: December 24, 1997
NIST Test No.: 844/xxxxxx-97/1

REPORT OF TEST
NIST Test # 39071S - Spectral Responsivity
Any Company

Manufacturer: UDT Sensors
Model #: UV100
Serial #: U1xxx

3. Results of Test

The absolute spectral responsivity in amperes per watt of the test photodiode is presented as a function of wavelength in table 1 and plotted in figure 1. The relative expanded uncertainties in the NIST absolute scale are described in Ref. [1]. The relative expanded uncertainty ($k = 2$) presented as a function of wavelength for this measurement is stated relative to absolute (SI) units and is listed in table 1 and is plotted in figure 1.

Table 2 lists the dimensions of the precision aperture furnished with the test photodiode. The reported aperture area and distance from the aperture plate to the photodiode surface do not enter into the responsivity measurement results because the optical beam underfills the aperture. The uncertainty values reported with the aperture dimensions are expanded uncertainties.

Figure 2a is a plot of the uniformity of the test photodiode, showing 0.2 % contours at 350 nm of the deviations from the responsivity at the photodiode center. Figure 2b is a 3-dimensional plot showing the responsivity relative to the center of the photodiode. Note that the response of the photodiode can vary by as much as a percent over the active area. This can lead to errors larger than the stated uncertainties if the irradiation geometry is significantly different from the test conditions described in section 2.

Figure 1

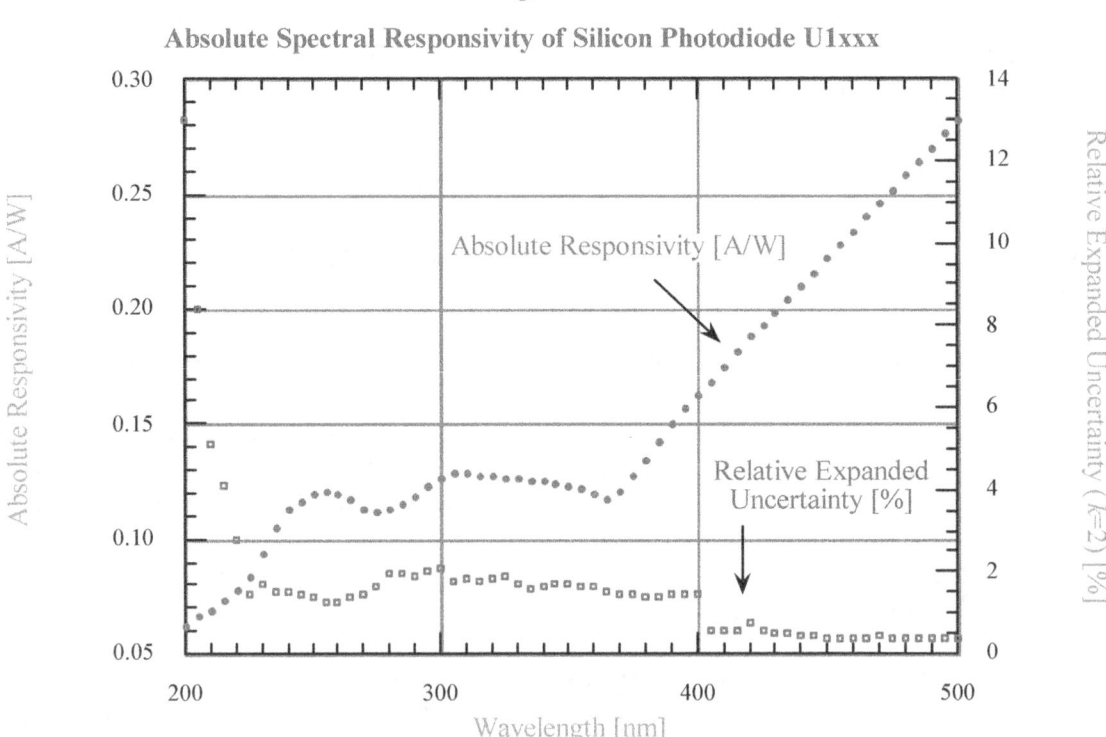

Absolute Spectral Responsivity of Silicon Photodiode U1xxx

Test Date: December 24, 1997
NIST Test No.: 844/xxxxxx-97/1

REPORT OF TEST
NIST Test # 39071S - Spectral Responsivity
Any Company

Manufacturer: UDT Sensors
Model #: UV100
Serial #: U1xxx

Table 1
Absolute Spectral Responsivity of Silicon Photodiode U1xxx

Wavelength [nm]	Absolute Responsivity [A/W]	Relative Expanded Uncertainty ($k=2$) [%]	Wavelength [nm]	Absolute Responsivity [A/W]	Relative Expanded Uncertainty ($k=2$) [%]
200	6.28E-2	13	350	1.24E-1	1.7
205	6.68E-2	8.4	355	1.22E-1	1.6
210	6.95E-2	5.1	360	1.20E-1	1.6
215	7.34E-2	4.1	365	1.18E-1	1.5
220	7.78E-2	2.8	370	1.21E-1	1.5
225	8.42E-2	1.5	375	1.27E-1	1.5
230	9.45E-2	1.7	380	1.35E-1	1.4
235	1.06E-1	1.5	385	1.43E-1	1.4
240	1.13E-1	1.5	390	1.50E-1	1.4
245	1.17E-1	1.4	395	1.57E-1	1.4
250	1.19E-1	1.4	400	1.63E-1	1.5
255	1.21E-1	1.3	405	1.689E-1	0.60
260	1.20E-1	1.3	410	1.750E-1	0.56
265	1.17E-1	1.4	415	1.816E-1	0.58
270	1.13E-1	1.5	420	1.882E-1	0.76
275	1.12E-1	1.7	425	1.935E-1	0.60
280	1.13E-1	2.0	430	1.993E-1	0.54
285	1.15E-1	2.0	435	2.049E-1	0.50
290	1.19E-1	1.9	440	2.100E-1	0.42
295	1.24E-1	2.0	445	2.160E-1	0.42
300	1.27E-1	2.1	450	2.220E-1	0.38
305	1.29E-1	1.8	455	2.281E-1	0.38
310	1.29E-1	1.8	460	2.341E-1	0.40
315	1.28E-1	1.8	465	2.401E-1	0.38
320	1.27E-1	1.8	470	2.462E-1	0.42
325	1.27E-1	1.9	475	2.523E-1	0.40
330	1.26E-1	1.7	480	2.583E-1	0.38
335	1.26E-1	1.6	485	2.643E-1	0.40
340	1.25E-1	1.6	490	2.704E-1	0.38
345	1.25E-1	1.7	495	2.764E-1	0.40
			500	2.824E-1	0.38

Table 2
Aperture Dimensions

Area: $0.5xxx \pm 0.0005$ cm^2

Distance from aperture plane to photodiode surface: 4.8 ± 0.4 mm

Test Date: December 24, 1997
NIST Test No.: 844/xxxxxx-97/1

REPORT OF TEST
NIST Test # 39071S - Spectral Responsivity
Any Company

Manufacturer: UDT Sensors
Model #: UV100
Serial #: U1xxx

Figure 2a
Responsivity Uniformity of Silicon Photodiode U1xxx
0.2 % contours at 350 nm; 1.5 mm resolution; 0.5 mm/Step

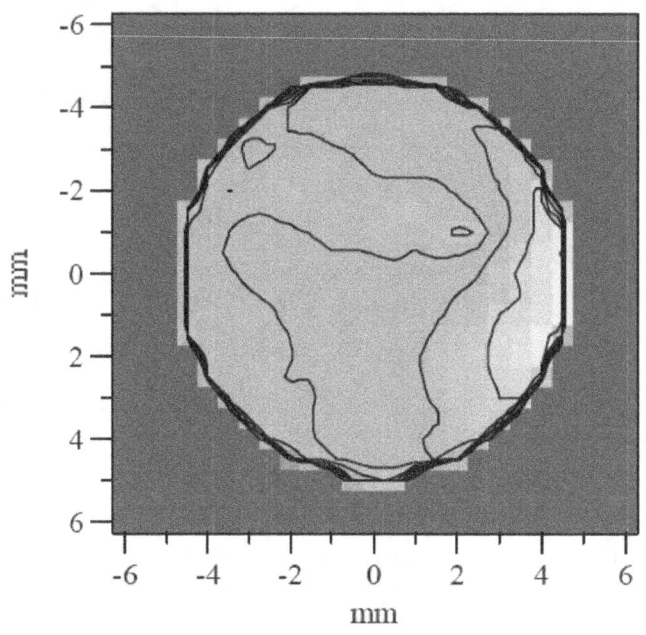

Figure 2b
Surface Plot of Responsivity Relative to
Center of Photodiode for Silicon Photodiode U1xxx
at 350 nm; 0.5 mm/Step

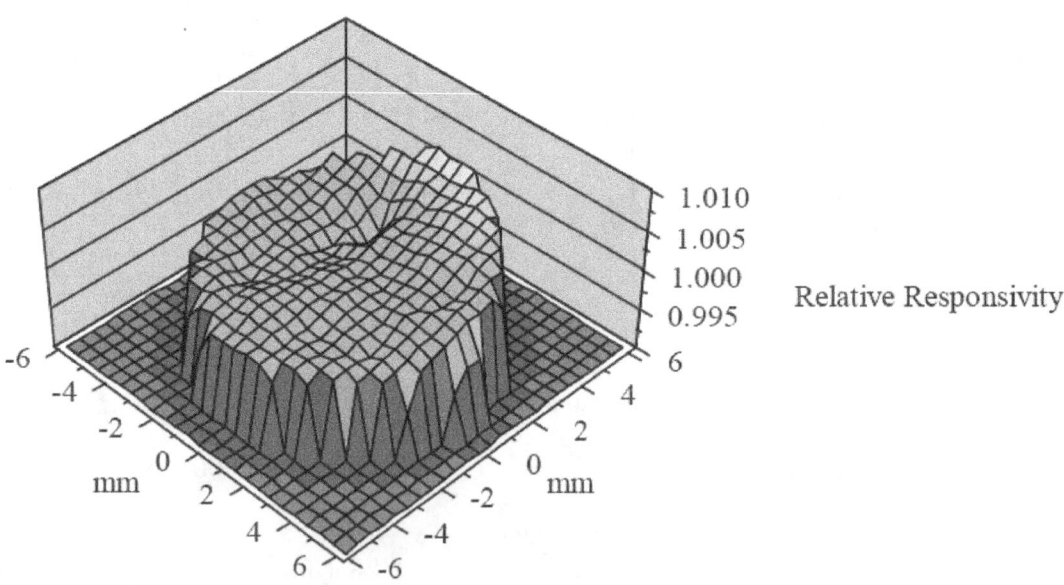

Test Date: December 24, 1997
NIST Test No.: 844/xxxxxx-97/1

REPORT OF TEST   Manufacturer: UDT Sensors
NIST Test # 39071S - Spectral Responsivity   Model #: UV100
Any Company   Serial #: U1xxx

4. General Information

The linearity of this type of photodiode is shown in figure 3 at 442 nm at irradiance levels from 0.1 mW/cm^2 to 1.1 mW/cm^2 with and without a reverse bias voltage. Each data point represents the ratio of the photodiode responsivity at the indicated irradiance to the responsivity within the linear region. The linearity was measured by using a beamsplitter to irradiate two photodiodes at approximately a 10:1 intensity ratio, with the diode aperture filled and uniformly irradiated. The linearity is dependent on the irradiation geometry and will differ for spot sizes significantly smaller than the aperture size.

The change in the responsivity for this type of photodiode at 442 nm is shown in figure 4 as a function of bias voltage. For wavelengths shorter than 450 nm, a 1 V bias can be used to improve the linearity of the photodiode without significantly changing the spectral responsivity. There will however be some leakage current which will limit the minimum usable signal.

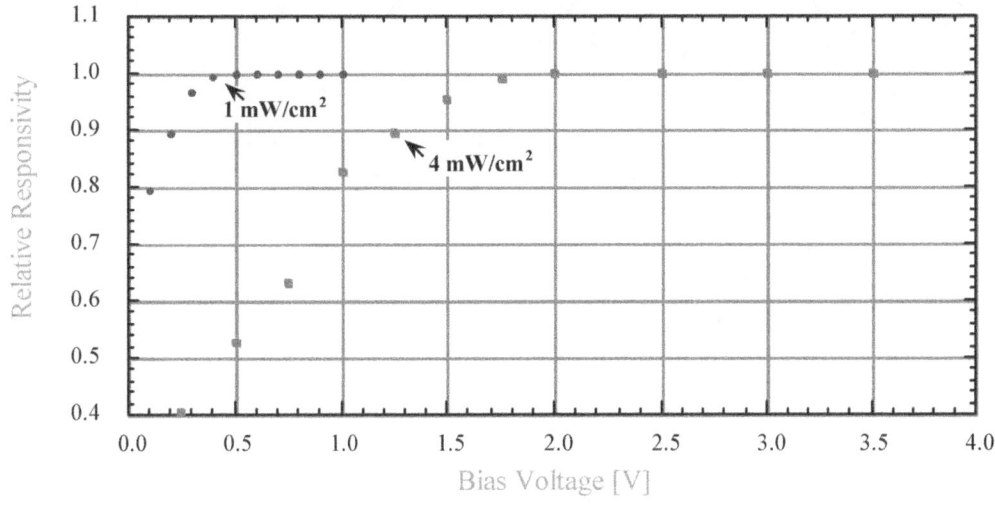

Figure 3
Linearity of UDT Sensors UV100 at 442 nm

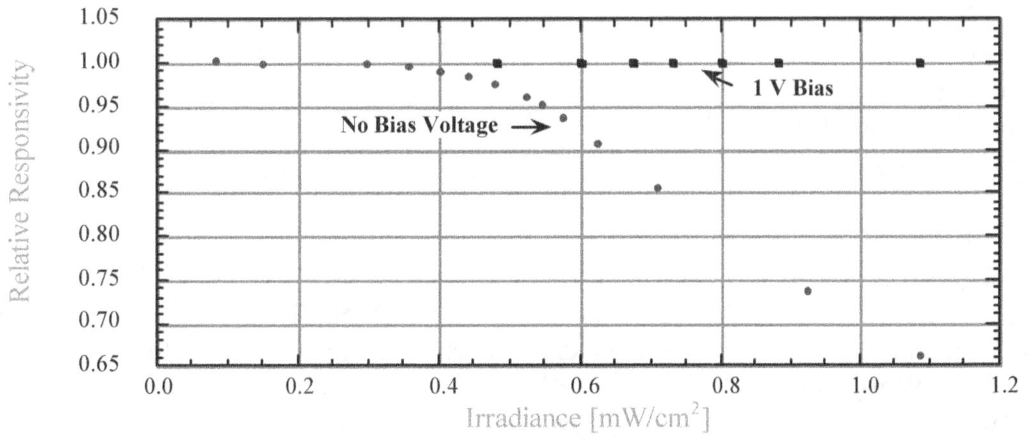

Figure 4
Responsivity dependence to bias voltage of UDT Sensors UV100 at 442 nm

Test Date: December 24, 1997
NIST Test No.: 844/xxxxxx-97/1

REPORT OF TEST
NIST Test # 39071S - Spectral Responsivity
Any Company

Manufacturer: UDT Sensors
Model #: UV100
Serial #: U1xxx

A documentation appendix includes operating instructions for the test photodiode. The laboratory temperature is reported for information only. It is not intended that this data be used for corrections to the spectral responsivity data in this report. This report shall not be reproduced, except in full, without the written approval of NIST.

Prepared by:

Sally S. Bruce
Optical Technology Division
Physics Laboratory
(301) 975-2323

Reviewed by:

Thomas C. Larason
Optical Technology Division
Physics Laboratory
(301) 975-2334

Approved by:

Joseph L. Dehmer
For the Director,
National Institute of
 Standards and Technology
(301) 975-2319

Reference:

[1] T. C. Larason, S. S. Bruce, and A. C. Parr, NIST Measurement Services: Spectroradiometric Detector Measurements: Part I - Ultraviolet Detectors and Part II - Visible to Near-Infrared Detectors, Natl. Inst. Stand. Technol., Spec. Publ. 250-41 (1998).

Test Date: December 24, 1997
NIST Test No.: 844/xxxxxx-97/1

REPORT OF TEST
NIST Test # 39071S - Spectral Responsivity
Any Company

Manufacturer: UDT Sensors
Model #: UV100
Serial #: U1xxx

APPENDIX:

OPERATING INSTRUCTIONS FOR NIST PHOTODIODE

The NIST characterized photodiode consists of a silicon photodiode with a removable precision aperture, a quartz window, and a BNC connector.

A. The photodiode should be rigidly mounted on a dual-axis tilt mount such that the photodiode can be tilted about two orthogonal axis. The photodiode should be adjusted to be perpendicular to the incident radiation.

B. The incident beam of radiation should be smaller than the aperture, and should be centered in the photodiode aperture.

C. The photodiode should be connected with a BNC cable to an electrometer grade amplifier (transimpedance amplifier) which measures the current from the photodiode.

D. The inside edge of the precision aperture is extremely delicate and should not be touched with fingers or any other object.

E. The diode window can be cleaned with lens tissue and spectral grade solvent. The precision aperture should be removed before cleaning the photodiode window.

Test Date: December 24, 1997
NIST Test No.: 844/xxxxxx-97/1

REPORT OF TEST

NIST Test # 39073S - Spectral Responsivity

for

Hamamatsu S2281 Silicon Photodiode, Cxxx

Submitted by:

Any Company
Mr. Daniel Doe
123 Calibration Court
Measurement City, MD 00000-0000

(See your Purchase Order No. XXXX-XX, dated January 1, 1993)

1. Description of Test Material

The test photodiode, labeled Cxxx, is a Hamamatsu S2281 silicon photodiode in an anodized aluminum mount with a removable precision aperture and a BNC connector. The active area of the photodiode is ≈ 1 cm^2.

2. Description of Test

The test photodiode was compared to two silicon photodiode working standards, H6xx and H6xx, using the NIST visible to near Infrared (Vis/NIR) monochromator-based comparator facility [1] from 350 nm to 1100 nm in 5 nm increments. The spectral comparisons between the test photodiode and working standard photodiodes were performed using a double monochromator and a quartz-halogen lamp as the tunable monochromatic source.

The circular exit aperture of the Vis/NIR monochromator was imaged ($\approx f/9$) on the test photodiode, resulting in a beam diameter at the photodiode of 1.1 mm. The beam was centered on, and underfilled, the aperture.

The wavelength scale of the monochromator was calibrated with several laser and emission lines and is accurate to $\pm$ 0.1 nm over the entire spectral range. The bandpass of the monochromator was 4 nm. The short-circuit photocurrent from the test photodiode and each working standard photodiode was measured with a calibrated transimpedance amplifier. The test photodiode and each working standard photodiode were measured with zero bias voltage. Beam power fluctuations were monitored with a beamsplitter and silicon photodiode. The absolute spectral responsivity scale is based on a high accuracy cryogenic radiometer with a relative expanded uncertainty ($k = 2$) to absolute (SI) units of 0.2 %.

The spatial uniformity of the responsivity across the test photodiode photosensitive area was measured at 500 nm using the described comparator facility. The uniformity was measured in 0.5 mm increments using a 1.1 mm diameter beam.

Laboratory Environment:
   Temperature: 23.x °C $\pm$ 0.3 °C

Test Date: December 24, 1997
NIST Test No.: 844/xxxxxx-97/2

REPORT OF TEST
NIST Test # 39073S - Spectral Responsivity
Any Company

Manufacturer: Hamamatsu
Model #: S2281
Serial #: Cxxx

3. Results of Test

The absolute spectral responsivity in amperes per watt of the test photodiode is presented as a function of wavelength in table 1 and is plotted in figure 1. The relative expanded uncertainties in the NIST absolute scale are described in Ref. [1]. The relative expanded uncertainty ($k = 2$) presented as a function of wavelength for this measurement is stated relative to absolute (SI) units and is listed in table 1 and plotted in figure 1.

Table 2 lists the dimensions of the precision aperture furnished with the test photodiode. The reported aperture area and distance from the aperture plate to the photodiode surface do not enter into the responsivity measurement results because the optical beam underfills the aperture. The uncertainty values reported with the aperture dimensions are expanded uncertainties.

Figure 2a is a plot of the uniformity of the test photodiode, showing 0.2 % contours at 500 nm of the deviations from the responsivity at the photodiode center. Figure 2b is a 3-dimensional plot showing the responsivity relative to the center of the photodiode. Errors larger than the stated uncertainties can occur if the irradiation geometry is significantly different from the test conditions described in section 2.

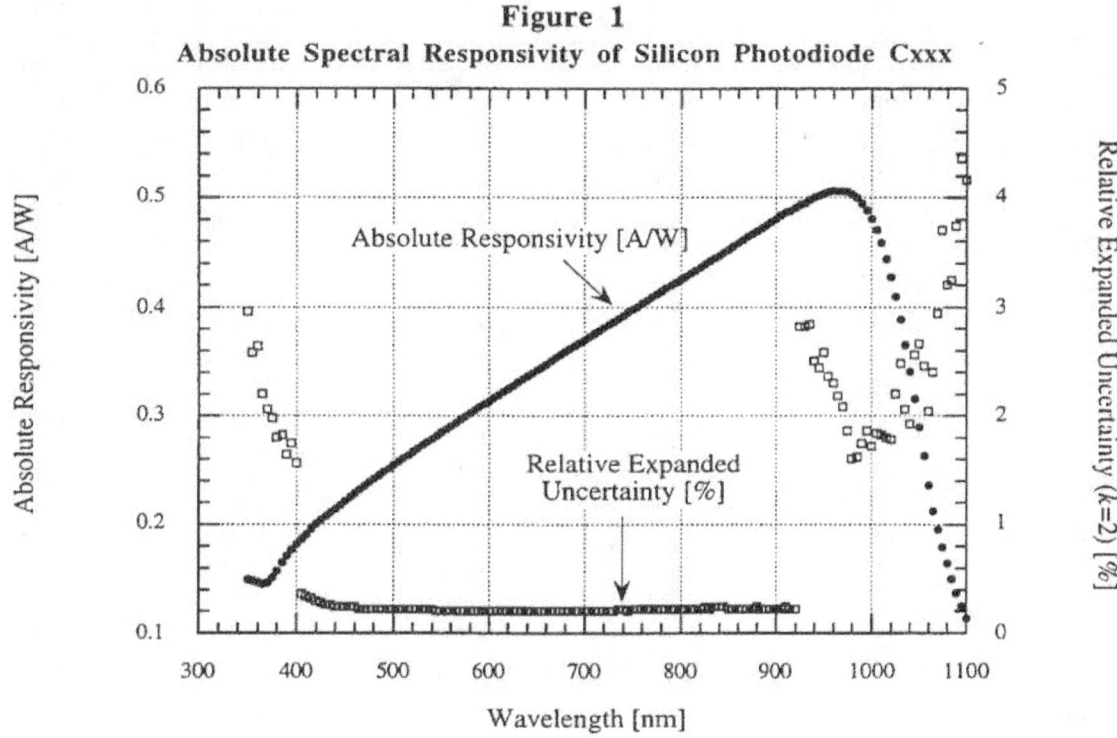

Figure 1
Absolute Spectral Responsivity of Silicon Photodiode Cxxx

Test Date: December 24, 1997
NIST Test No.: 844/xxxxxx-97/2

REPORT OF TEST
NIST Test # 39073S - Spectral Responsivity
Any Company

Manufacturer: Hamamatsu
Model #: S2281
Serial #: Cxxx

Table 1
Absolute Spectral Responsivity of Silicon Photodiode Cxxx

Wavelength [nm]	Absolute Responsivity [A/W]	Relative Expanded Uncertainty ($k = 2$) [%]	Wavelength [nm]	Absolute Responsivity [A/W]	Relative Expanded Uncertainty ($k = 2$) [%]
350	1.50E-1	3.0	550	2.848E-1	0.20
355	1.49E-1	2.6	555	2.877E-1	0.20
360	1.47E-1	2.6	560	2.906E-1	0.20
365	1.45E-1	2.2	565	2.935E-1	0.20
370	1.47E-1	2.1	570	2.965E-1	0.20
375	1.51E-1	2.0	575	2.994E-1	0.20
380	1.58E-1	1.8	580	3.023E-1	0.20
385	1.65E-1	1.8	585	3.052E-1	0.20
390	1.71E-1	1.6	590	3.081E-1	0.20
395	1.77E-1	1.7	595	3.109E-1	0.20
400	1.82E-1	1.6	600	3.138E-1	0.20
405	1.860E-1	0.36	605	3.166E-1	0.20
410	1.909E-1	0.34	610	3.195E-1	0.20
415	1.959E-1	0.32	615	3.223E-1	0.20
420	2.008E-1	0.30	620	3.252E-1	0.20
425	2.041E-1	0.28	625	3.280E-1	0.20
430	2.079E-1	0.26	630	3.308E-1	0.20
435	2.113E-1	0.26	635	3.337E-1	0.20
440	2.141E-1	0.24	640	3.365E-1	0.20
445	2.178E-1	0.24	645	3.393E-1	0.20
450	2.214E-1	0.24	650	3.421E-1	0.20
455	2.250E-1	0.24	655	3.449E-1	0.20
460	2.284E-1	0.24	660	3.477E-1	0.20
465	2.319E-1	0.22	665	3.505E-1	0.20
470	2.352E-1	0.22	670	3.534E-1	0.20
475	2.385E-1	0.22	675	3.562E-1	0.20
480	2.418E-1	0.22	680	3.589E-1	0.20
485	2.450E-1	0.22	685	3.617E-1	0.20
490	2.482E-1	0.22	690	3.645E-1	0.20
495	2.513E-1	0.22	695	3.673E-1	0.20
500	2.545E-1	0.22	700	3.701E-1	0.20
505	2.576E-1	0.22	705	3.729E-1	0.20
510	2.607E-1	0.22	710	3.756E-1	0.20
515	2.638E-1	0.22	715	3.784E-1	0.20
520	2.668E-1	0.22	720	3.812E-1	0.20
525	2.699E-1	0.22	725	3.840E-1	0.20
530	2.729E-1	0.22	730	3.868E-1	0.20
535	2.759E-1	0.22	735	3.896E-1	0.22
540	2.788E-1	0.22	740	3.923E-1	0.22
545	2.818E-1	0.20	745	3.951E-1	0.20

Test Date: December 24, 1997
NIST Test No.: 844/xxxxxx-97/2

REPORT OF TEST
NIST Test # 39073S - Spectral Responsivity
Any Company

Manufacturer: Hamamatsu
Model #: S2281
Serial #: Cxxx

Table 1 (cont.)
Absolute Spectral Responsivity of Silicon Photodiode Cxxx

Wavelength [nm]	Absolute Responsivity [A/W]	Relative Expanded Uncertainty ($k = 2$) [%]	Wavelength [nm]	Absolute Responsivity [A/W]	Relative Expanded Uncertainty ($k = 2$) [%]
750	3.978E-1	0.22	925	4.93E-1	2.8
755	4.007E-1	0.22	930	4.95E-1	2.8
760	4.034E-1	0.22	935	4.97E-1	2.8
765	4.061E-1	0.22	940	5.00E-1	2.5
770	4.088E-1	0.22	945	5.02E-1	2.4
775	4.116E-1	0.22	950	5.03E-1	2.6
780	4.144E-1	0.22	955	5.04E-1	2.4
785	4.171E-1	0.22	960	5.06E-1	2.3
790	4.199E-1	0.22	965	5.05E-1	2.2
795	4.227E-1	0.22	970	5.05E-1	2.1
800	4.254E-1	0.22	975	5.04E-1	1.9
805	4.282E-1	0.22	980	5.02E-1	1.6
810	4.309E-1	0.22	985	4.98E-1	1.6
815	4.336E-1	0.22	990	4.93E-1	1.7
820	4.364E-1	0.22	995	4.87E-1	1.9
825	4.391E-1	0.24	1000	4.79E-1	1.7
830	4.418E-1	0.22	1005	4.69E-1	1.8
835	4.446E-1	0.24	1010	4.57E-1	1.8
840	4.473E-1	0.24	1015	4.42E-1	1.8
845	4.502E-1	0.24	1020	4.26E-1	1.8
850	4.528E-1	0.22	1025	4.08E-1	2.2
855	4.556E-1	0.22	1030	3.87E-1	2.5
860	4.584E-1	0.22	1035	3.63E-1	2.1
865	4.610E-1	0.22	1040	3.38E-1	1.9
870	4.638E-1	0.22	1045	3.13E-1	2.6
875	4.665E-1	0.22	1050	2.86E-1	2.7
880	4.693E-1	0.24	1055	2.60E-1	2.5
885	4.720E-1	0.22	1060	2.33E-1	2.0
890	4.747E-1	0.22	1065	2.09E-1	2.4
895	4.774E-1	0.22	1070	1.91E-1	2.9
900	4.802E-1	0.22	1075	1.75E-1	3.7
905	4.833E-1	0.22	1080	1.60E-1	3.2
910	4.861E-1	0.24	1085	1.46E-1	3.2
915	4.876E-1	0.22	1090	1.33E-1	3.7
920	4.902E-1	0.22	1095	1.20E-1	4.4
			1100	1.09E-1	4.2

Test Date: December 24, 1997
NIST Test No.: 844/xxxxxx-97/2

REPORT OF TEST
NIST Test # 39073S - Spectral Responsivity
Any Company

Manufacturer: Hamamatsu
Model #: S2281
Serial #: Cxxx

Figure 2a
Responsivity Uniformity of Silicon Photodiode Cxxx
0.2 % contours at 500 nm; 1.1 mm resolution; 0.5 mm/Step

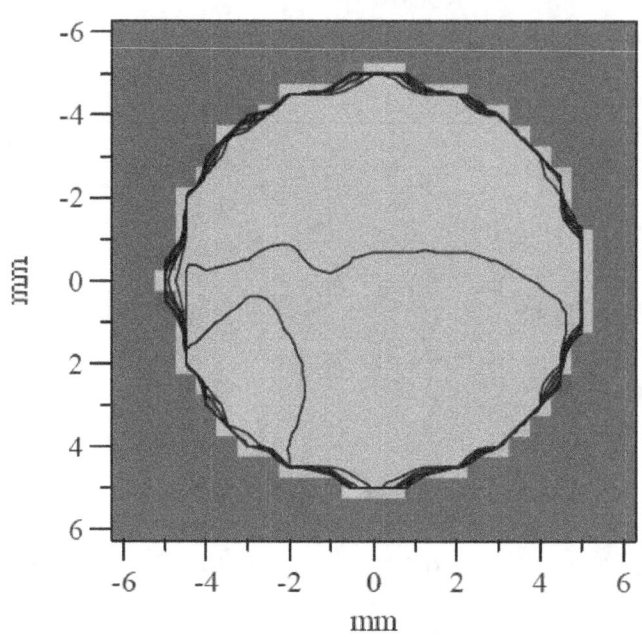

Figure 2b
Surface Plot of Responsivity Relative to
Center of Photodiode for Silicon Photodiode Cxxx
at 500 nm; 0.5 mm/Step

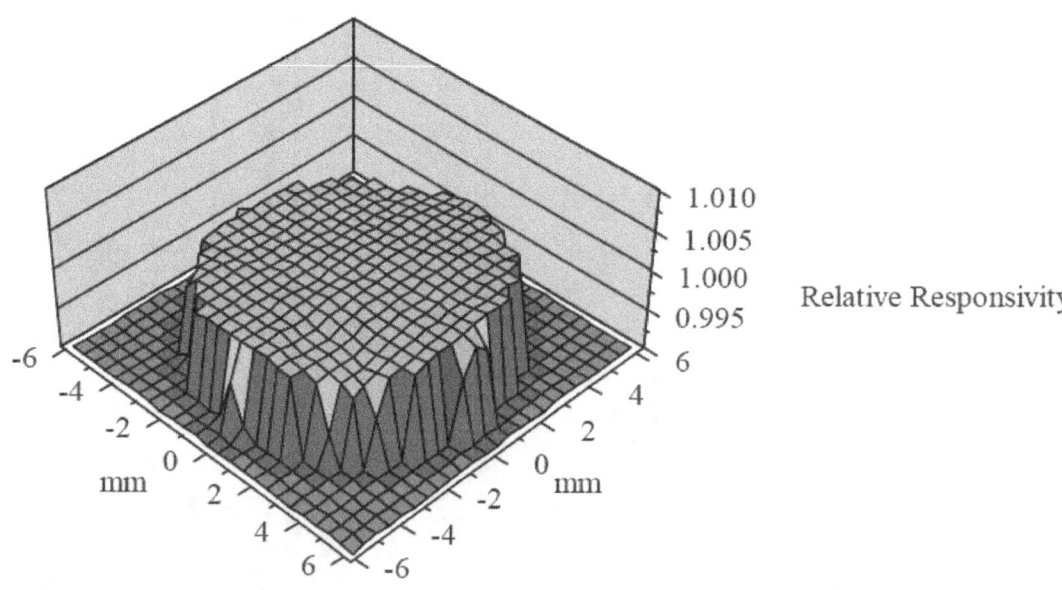

Test Date: December 24, 1997
NIST Test No.: 844/xxxxxx-97/2

REPORT OF TEST
NIST Test # 39073S - Spectral Responsivity
Any Company

Manufacturer: Hamamatsu
Model #: S2281
Serial #: Cxxx

Table 2
Aperture Dimensions

Area: $0.5\text{xxx} \pm 0.0005 \text{ cm}^2$

Distance from aperture plane
to photodiode surface: 5.0 ± 0.5 mm

4. General Information

A documentation appendix includes operating instructions for the test photodiode. The laboratory temperature is reported for information only. It is not intended that this data be used for corrections to the spectral responsivity data in this report. This report shall not be reproduced, except in full, without the written approval of NIST.

Prepared by: Reviewed by:

Sally S. Bruce   Thomas C. Larason
Optical Technology Division   Optical Technology Division
Physics Laboratory   Physics Laboratory
(301) 975-2323   (301) 975-2334

Approved by:

Joseph L. Dehmer
For the Director,
National Institute of
 Standards and Technology
(301) 975-2319

Reference:

[1]   T. C. Larason, S. S. Bruce, and A. C. Parr, NIST Measurement Services: Spectroradiometric Detector Measurements: Part I - Ultraviolet Detectors and Part II - Visible to Near-Infrared Detectors, Natl. Inst. Stand. Technol., Spec. Publ. 250-41 (1998).

REPORT OF TEST
NIST Test # 39073S - Spectral Responsivity
Any Company

Manufacturer: Hamamatsu
Model #: S2281
Serial #: Cxxx

APPENDIX:

OPERATING INSTRUCTIONS
FOR
NIST PHOTODIODE

The NIST characterized photodiode consists of a silicon photodiode with a removable precision aperture, a quartz window, and a BNC connector.

A. The photodiode should be rigidly mounted on a dual-axis tilt mount such that the photodiode can be tilted about two orthogonal axis. The photodiode should be adjusted to be perpendicular to the incident radiation.

B. The incident beam of radiation should be smaller than the aperture, and should be centered in the photodiode aperture.

C. The photodiode should be connected with a BNC cable to an electrometer grade amplifier (transimpedance amplifier) which measures the current from the photodiode.

D. The inside edge of the precision aperture is extremely delicate and should not be touched with fingers or any other object.

E. The diode window can be cleaned with lens tissue and spectral grade solvent. The precision aperture should be removed before cleaning the photodiode window.

REPORT OF TEST

NIST Test # 39075S - Spectral Responsivity

for

Acme Instruments Germanium Photodiode
Model xx, S/N yyy

Submitted by:

Any Company
Mr. Daniel Doe
123 Calibration Court
Measurement City, MD 00000-0000

(See your Purchase Order No. XXXX-XX, dated January 1, 1993)

1. Description of Test Material

The test photodiode, Acme Instruments model xx, S/N yyy, consists of a thermoelectrically cooled germanium photodiode mounted in a cylindrical aluminum housing with the output signal available on a BNC connector. The active area of the photodiode is $\approx$ x cm^2.

The test photodiode was measured as supplied by Any Company. The test photodiode was also supplied with an Acme Instruments Thermoelectric Cooler Controller model zz, S/N zzz.

2. Description of Test

The test photodiode was compared to two cooled germanium photodiode working standards, Ge #x and Ge #x, using the NIST visible to near Infrared (Vis/NIR) monochromator-based comparator facility [1] from 700 nm to 1800 nm in 5 nm increments. The spectral comparisons between the test photodiode and working standard photodiodes were performed using a double monochromator and a quartz-halogen lamp as the tunable monochromatic source.

The circular exit aperture of the Vis/NIR monochromator was imaged ($\approx f/9$) on the test photodiode, resulting in a beam diameter at the photosensitive area of 1.1 mm. The beam was centered on, and underfilled, the photosensitive area. The photocurrent was measured with the thermoelectric cooler controller set at -30 °C.

The wavelength scale of the monochromator was calibrated with several laser and emission lines and is accurate to ± 0.1 nm over the entire spectral range. The bandpass of the monochromator was 4 nm. The short-circuit photocurrent from the test photodiode and each working standard photodiode was measured with a calibrated transimpedance amplifier. The test photodiode and each working standard photodiode were measured with zero bias voltage. Beam power fluctuations were monitored with a beamsplitter and cooled germanium photodiode. The absolute

Laboratory Environment:
    Temperature: 23.x °C ± 0.3 °C

Test Date: December 24, 1997
NIST Test No.: 844/xxxxxx-97/3

REPORT OF TEST
NIST Test # 39075S - Spectral Responsivity
Any Company

Manufacturer: Acme Instruments
Model #: xx
Serial #: yyy

spectral responsivity scale is based on a high accuracy cryogenic radiometer, with a relative expanded uncertainty ($k = 2$) to absolute (SI) units of 0.2 %.

3. Results of Test

The absolute spectral responsivity in amperes per watt of the test photodiode is presented as a function of wavelength in table 1 and is plotted in figure 1. The relative expanded uncertainty in the NIST absolute scale is described in Ref. [1]. The relative expanded uncertainty ($k = 2$) presented as a function of wavelength for this measurement is stated relative to absolute (SI) units and is listed in table 1 and plotted in figure 1.

The reported relative expanded uncertainty does not include estimates for several components that are unknown for this test photodiode. The unknown uncertainty components are photodiode responsivity uniformity, polarization sensitivity, linearity, temperature coefficient, and long-term stability. These components could significantly add to the reported uncertainty. Errors larger than the stated uncertainties can occur if the irradiation geometry is significantly different from the test conditions described in section 2.

Figure 1

Absolute Spectral Responsivity of Acme Instruments Photodiode Model xx (S/N yyy)

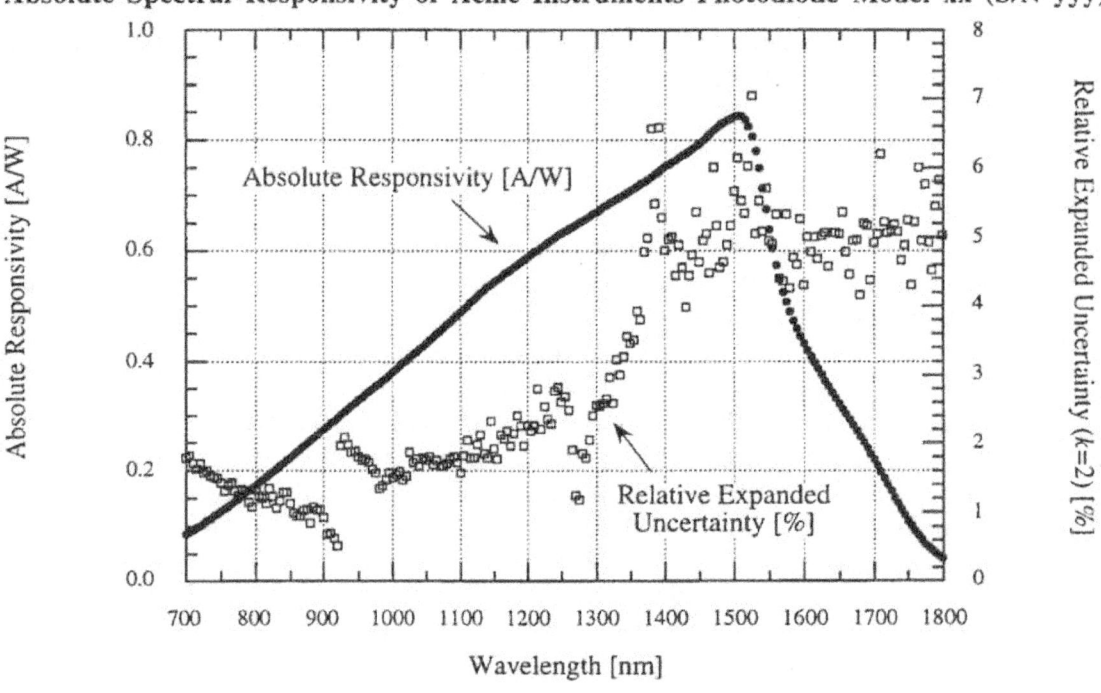

Test Date: December 24, 1997
NIST Test No.: 844/xxxxxx-97/3

REPORT OF TEST
NIST Test # 39075S - Spectral Responsivity
Any Company

Manufacturer: Acme Instruments
Model #: xx
Serial #: yyy

Table 1
Absolute Spectral Responsivity of Acme Instruments Model xx (S/N yyy)

Wavelength [nm]	Absolute Responsivity [A/W]	Relative Expanded Uncertainty ($k=2$) [%]	Wavelength [nm]	Absolute Responsivity [A/W]	Relative Expanded Uncertainty ($k=2$) [%]
700	8.55E-2	1.8	900	2.758E-1	0.92
705	8.92E-2	1.8	905	2.810E-1	0.68
710	9.32E-2	1.7	910	2.861E-1	0.70
715	9.72E-2	1.6	915	2.914E-1	0.62
720	1.01E-1	1.7	920	2.968E-1	0.52
725	1.05E-1	1.6	925	3.02E-1	2.0
730	1.09E-1	1.6	930	3.08E-1	2.1
735	1.14E-1	1.5	935	3.13E-1	2.0
740	1.18E-1	1.5	940	3.18E-1	1.9
745	1.22E-1	1.5	945	3.24E-1	1.9
750	1.27E-1	1.4	950	3.29E-1	1.8
755	1.31E-1	1.3	955	3.34E-1	1.7
760	1.36E-1	1.4	960	3.40E-1	1.8
765	1.40E-1	1.4	965	3.45E-1	1.7
770	1.45E-1	1.3	970	3.50E-1	1.6
775	1.50E-1	1.2	975	3.54E-1	1.6
780	1.54E-1	1.3	980	3.59E-1	1.3
785	1.59E-1	1.3	985	3.64E-1	1.4
790	1.64E-1	1.1	990	3.69E-1	1.5
795	1.69E-1	1.1	995	3.75E-1	1.6
800	1.73E-1	1.3	1000	3.81E-1	1.5
805	1.78E-1	1.2	1005	3.86E-1	1.5
810	1.83E-1	1.2	1010	3.92E-1	1.6
815	1.88E-1	1.1	1015	3.97E-1	1.5
820	1.93E-1	1.3	1020	4.02E-1	1.5
825	1.98E-1	1.2	1025	4.07E-1	1.9
830	2.03E-1	1.1	1030	4.12E-1	1.7
835	2.08E-1	1.2	1035	4.16E-1	1.8
840	2.13E-1	1.3	1040	4.21E-1	1.7
845	2.19E-1	1.3	1045	4.27E-1	1.8
850	2.24E-1	1.1	1050	4.32E-1	1.7
855	2.289E-1	0.98	1055	4.38E-1	1.8
860	2.341E-1	0.94	1060	4.43E-1	1.7
865	2.393E-1	0.94	1065	4.49E-1	1.7
870	2.44E-1	1.0	1070	4.55E-1	1.7
875	2.50E-1	1.0	1075	4.60E-1	1.7
880	2.549E-1	0.84	1080	4.66E-1	1.7
885	2.60E-1	1.1	1085	4.71E-1	1.8
890	2.65E-1	1.0	1090	4.77E-1	1.8
895	2.71E-1	1.0	1095	4.82E-1	1.7

Test Date: December 24, 1997
NIST Test No.: 844/xxxxxx-97/3

REPORT OF TEST
NIST Test # 39075S - Spectral Responsivity
Any Company

Manufacturer: Acme Instruments
Model #: xx
Serial #: yyy

Table 1 (cont.)
Absolute Spectral Responsivity of Acme Instruments Model xx (S/N yyy)

Wavelength [nm]	Absolute Responsivity [A/W]	Relative Expanded Uncertainty ($k = 2$) [%]	Wavelength [nm]	Absolute Responsivity [A/W]	Relative Expanded Uncertainty ($k = 2$) [%]
1100	4.88E-1	1.6	1300	6.70E-1	2.6
1105	4.94E-1	1.8	1305	6.75E-1	2.5
1110	5.00E-1	2.0	1310	6.79E-1	2.6
1115	5.06E-1	1.8	1315	6.83E-1	2.6
1120	5.12E-1	1.8	1320	6.86E-1	3.0
1125	5.18E-1	2.0	1325	6.90E-1	2.6
1130	5.23E-1	2.1	1330	6.94E-1	3.2
1135	5.29E-1	1.8	1335	6.97E-1	3.0
1140	5.34E-1	1.8	1340	7.01E-1	3.3
1145	5.39E-1	2.3	1345	7.05E-1	3.6
1150	5.44E-1	1.9	1350	7.09E-1	3.5
1155	5.49E-1	1.8	1355	7.12E-1	3.5
1160	5.53E-1	2.1	1360	7.16E-1	3.9
1165	5.58E-1	2.1	1365	7.20E-1	3.8
1170	5.62E-1	2.2	1370	7.24E-1	4.8
1175	5.67E-1	2.0	1375	7.29E-1	5.0
1180	5.71E-1	2.1	1380	7.33E-1	6.6
1185	5.75E-1	2.4	1385	7.38E-1	5.5
1190	5.80E-1	2.3	1390	7.43E-1	6.6
1195	5.84E-1	2.0	1395	7.47E-1	5.3
1200	5.89E-1	2.3	1400	7.52E-1	4.8
1205	5.93E-1	2.2	1405	7.56E-1	5.0
1210	5.98E-1	2.3	1410	7.60E-1	5.0
1215	6.02E-1	2.8	1415	7.64E-1	4.4
1220	6.06E-1	2.2	1420	7.68E-1	4.9
1225	6.10E-1	2.5	1425	7.72E-1	4.6
1230	6.15E-1	2.4	1430	7.76E-1	4.0
1235	6.19E-1	2.3	1435	7.81E-1	4.4
1240	6.23E-1	2.8	1440	7.85E-1	4.7
1245	6.28E-1	2.8	1445	7.90E-1	5.4
1250	6.32E-1	2.6	1450	7.94E-1	4.6
1255	6.36E-1	2.7	1455	8.00E-1	4.9
1260	6.40E-1	2.5	1460	8.05E-1	5.0
1265	6.43E-1	1.9	1465	8.11E-1	4.5
1270	6.47E-1	1.2	1470	8.17E-1	6.0
1275	6.50E-1	1.2	1475	8.23E-1	5.2
1280	6.54E-1	1.8	1480	8.28E-1	4.6
1285	6.58E-1	1.8	1485	8.32E-1	4.6
1290	6.62E-1	2.0	1490	8.36E-1	4.9
1295	6.66E-1	2.4	1495	8.40E-1	5.2

Test Date: December 24, 1997
NIST Test No.: 844/xxxxxx-97/3

REPORT OF TEST
NIST Test # 39075S - Spectral Responsivity
Any Company

Manufacturer: Acme Instruments
Model #: xx
Serial #: yyy

Table 1 (cont.)
Absolute Spectral Responsivity of Acme Instruments Model xx (S/N yyy)

Wavelength [nm]	Absolute Responsivity [A/W]	Relative Expanded Uncertainty ($k=2$) [%]	Wavelength [nm]	Absolute Responsivity [A/W]	Relative Expanded Uncertainty ($k=2$) [%]
1500	8.43E-1	5.7	1650	3.23E-1	5.0
1505	8.44E-1	6.1	1655	3.13E-1	5.4
1510	8.42E-1	5.5	1660	3.03E-1	4.8
1515	8.37E-1	5.3	1665	2.93E-1	4.5
1520	8.25E-1	6.0	1670	2.83E-1	4.9
1525	8.06E-1	7.0	1675	2.73E-1	5.0
1530	7.80E-1	5.0	1680	2.62E-1	4.2
1535	7.49E-1	5.5	1685	2.52E-1	5.2
1540	7.13E-1	5.1	1690	2.41E-1	5.2
1545	6.76E-1	5.7	1695	2.31E-1	4.4
1550	6.39E-1	4.9	1700	2.20E-1	4.9
1555	6.05E-1	4.9	1705	2.09E-1	5.0
1560	5.74E-1	5.3	1710	1.98E-1	6.2
1565	5.48E-1	4.4	1715	1.87E-1	5.2
1570	5.26E-1	4.4	1720	1.75E-1	5.1
1575	5.07E-1	5.3	1725	1.64E-1	5.1
1580	4.90E-1	4.3	1730	1.52E-1	5.2
1585	4.74E-1	4.7	1735	1.41E-1	5.1
1590	4.59E-1	4.6	1740	1.30E-1	4.7
1595	4.45E-1	5.3	1745	1.20E-1	4.9
1600	4.32E-1	4.3	1750	1.10E-1	5.2
1605	4.20E-1	5.0	1755	1.00E-1	4.3
1610	4.08E-1	4.8	1760	9.14E-2	5.2
1615	3.97E-1	5.0	1765	8.31E-2	6.0
1620	3.87E-1	4.7	1770	7.55E-2	4.9
1625	3.76E-1	5.0	1775	6.86E-2	5.8
1630	3.65E-1	5.1	1780	6.23E-2	4.9
1635	3.54E-1	4.6	1785	5.66E-2	4.5
1640	3.43E-1	5.1	1790	5.14E-2	5.4
1645	3.33E-1	5.1	1795	4.65E-2	5.8
			1800	4.21E-2	5.0

Test Date: December 24, 1997
NIST Test No.: 844/xxxxxx-97/3

REPORT OF TEST
NIST Test # 39075S - Spectral Responsivity
Any Company

Manufacturer: Acme Instruments
Model #: xx
Serial #: yyy

4. General Information

The laboratory temperature is reported for information only. It is not intended that this data be used for corrections to the spectral responsivity data in this report. This report shall not be reproduced, except in full, without the written approval of NIST.

Prepared by:

Reviewed by:

Sally S. Bruce
Optical Technology Division
Physics Laboratory
(301) 975-2323

Thomas C. Larason
Optical Technology Division
Physics Laboratory
(301) 975-2334

Approved by:

Joseph L. Dehmer
For the Director,
National Institute of
 Standards and Technology
(301) 975-2319

Reference:

[1] T. C. Larason, S. S. Bruce, and A. C. Parr, NIST Measurement Services: Spectroradiometric Detector Measurements: Part I - Ultraviolet Detectors and Part II - Visible to Near-Infrared Detectors, Natl. Inst. Stand. Technol., Spec. Publ. 250-41 (1998).

REPORT OF TEST

NIST Test # 39081S - Responsivity Spatial Uniformity

for

Acme Instruments Silicon Photodiode
Model aa, S/N bbb

Submitted by:

Dr. Matthew Metrologist
Bigtime Government National Laboratory
Group PPL2 (Mail Stop: F769)
Secretcity, NM 87545-0001

(See your Purchase Order No. YY-YYYY, dated April 1, 1997)

1. Description of Test Material

The test photodiode, Acme Instruments model aa, S/N bbb, consists of a silicon photodiode mounted in a cylindrical aluminum housing with the output signal available on a BNC connector. The active area of the photodiode is ≈ 1 cm^2. The test photodiode was measured as supplied by Bigtime Government National Laboratory.

2. Description of Test

The relative spatial uniformity of the responsivity across the test photodiode photosensitive area was measured using the NIST visible to near Infrared (Vis/NIR) monochromator-based comparator facility [1] from 400 nm to 700 nm in 100 nm increments. The relative uniformity was measured in 0.5 mm increments with a 1.1 mm diameter beam using a double monochromator and a quartz-halogen lamp as the tunable monochromatic source. The circular exit aperture of the Vis/NIR monochromator was imaged ($\approx f/9$) onto the photosensitive area resulting in a beam diameter of 1.1 mm.

The wavelength scale of the monochromator was calibrated with several laser and emission lines and is accurate to ± 0.1 nm over the entire spectral range. The bandpass of the monochromator was 4 nm. The short-circuit photocurrent from the test photodiode was measured with a calibrated transimpedance amplifier. The test photodiode was measured with zero bias voltage. Beam power fluctuations were monitored with a beamsplitter and silicon photodiode.

3. Results of Test

Figure 1a is a plot of the relative uniformity of the test photodiode photosensitive area, showing 0.2 % contours at 400 nm of the deviations from the responsivity at the center of the photosensitive area. Figure 1b is a 3-dimensional plot showing the responsivity relative to the center of the photosensitive area.

Laboratory Environment:
    Temperature: 23.x °C ± 0.3 °C

Test Date: December 24, 1997
NIST Test No.: 844/yyyyyy-97

REPORT OF TEST
NIST Test # 39081S - Responsivity Spatial Uniformity
Bigtime Government National Laboratory

Manufacturer: Acme Instruments
Model #: aa
Serial #: bbb

Figures 2, 3, and 4 are similar plots of the relative uniformity at 500 nm, 600 nm, and 700 nm respectively.

The relative expanded uncertainty ($k = 2$) for the responsivity values is 0.0024 %. This is the repeatability of the measured relative responsivity in the central portion of the active area during the measurement scan. The measurement repeatability uncertainty depends on the SNR of the detector and can vary spectrally. Note that the variation in responsivity over the measured area is much larger than this uncertainty value. The reported uncertainty is not an indication of the uniformity measurement reproducibility. The uncertainty analysis is described in Ref. [1].

The intended primary use of the reported uniformity results is qualitative. That is, to indicate if any large discontinuities are present in the responsivity uniformity which can lead to larger than expected uncertainties in absolute responsivity measurements. Quantitative application of the reported uniformity results requires examination of the irradiance geometry and equipment involved. The generalized application of the uniformity measurement results is currently being studied.

4. General Information

The laboratory temperature is reported for information only. It is not intended that this data be used for corrections to the spectral responsivity data in this report. This report shall not be reproduced, except in full, without the written approval of NIST.

Prepared by:

Reviewed by:

Sally S. Bruce
Optical Technology Division
Physics Laboratory
(301) 975-2323

Thomas C. Larason
Optical Technology Division
Physics Laboratory
(301) 975-2334

Approved by:

Joseph L. Dehmer
For the Director,
National Institute of
 Standards and Technology
(301) 975-2319

Reference:

[1] T. C. Larason, S. S. Bruce, and A. C. Parr, NIST Measurement Services: Spectroradiometric Detector Measurements: Part I - Ultraviolet Detectors and Part II - Visible to Near-Infrared Detectors, Natl. Inst. Stand. Technol., Spec. Publ. 250-41 (1998).

Test Date: December 24, 1997
NIST Test No.: 844/yyyyyy-97

REPORT OF TEST
NIST Test # 39081S - Responsivity Spatial Uniformity
Bigtime Government National Laboratory

Manufacturer: Acme Instruments
Model #: aa
Serial #: bbb

Figure 1a
Responsivity Uniformity of Acme Instruments aa, S/N bbb
0.2 % contours at 400 nm; 1.1 mm resolution; 0.5 mm/Step

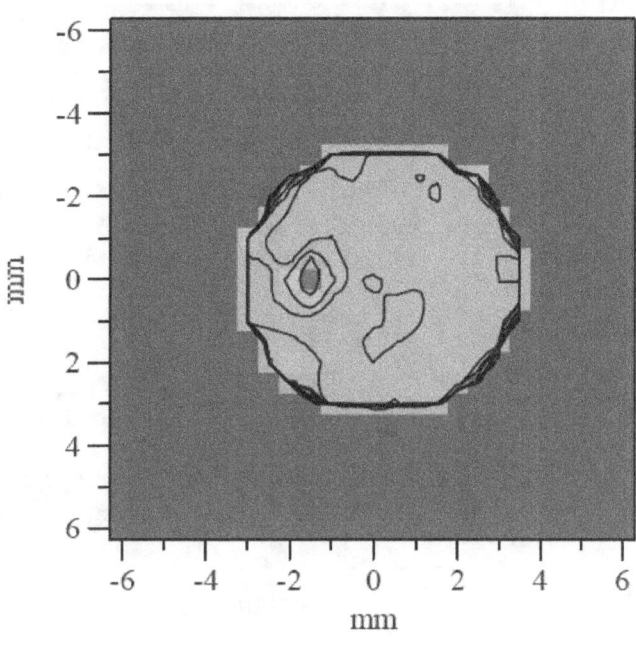

Figure 1b
Surface Plot of Responsivity Relative to
Center of Photosensitive Area for Acme Instruments aa, S/N bbb
at 400 nm; 0.5 mm/Step

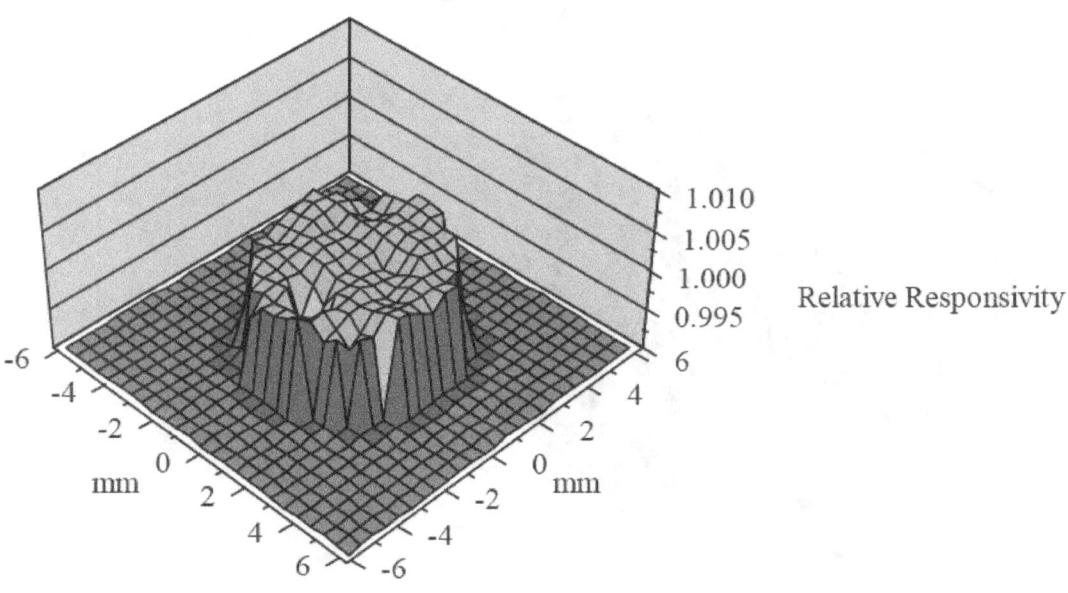

Test Date: December 24, 1997
NIST Test No.: 844/yyyyyy-97

REPORT OF TEST
NIST Test # 39081S - Responsivity Spatial Uniformity
Bigtime Government National Laboratory

Manufacturer: Acme Instruments
Model #: aa
Serial #: bbb

Figure 2a
Responsivity Uniformity of Acme Instruments aa, S/N bbb
0.2 % contours at 500 nm; 1.1 mm resolution; 0.5 mm/Step

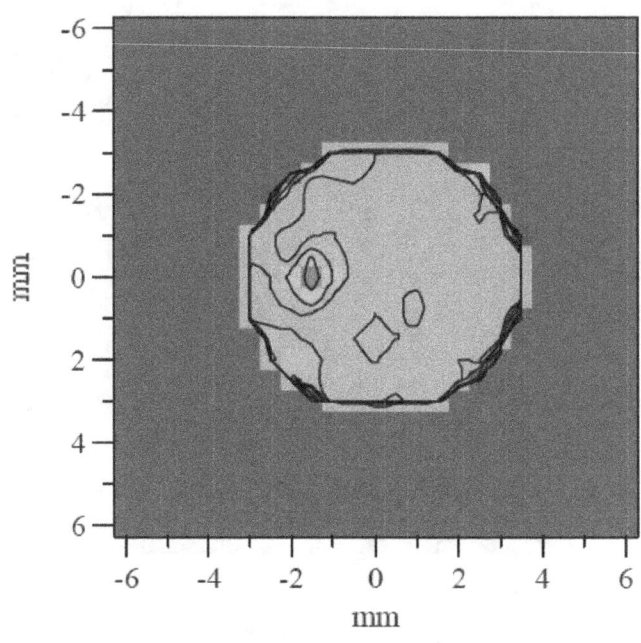

Figure 2b
Surface Plot of Responsivity Relative to
Center of Photosensitive Area for Acme Instruments aa, S/N bbb
at 500 nm; 0.5 mm/Step

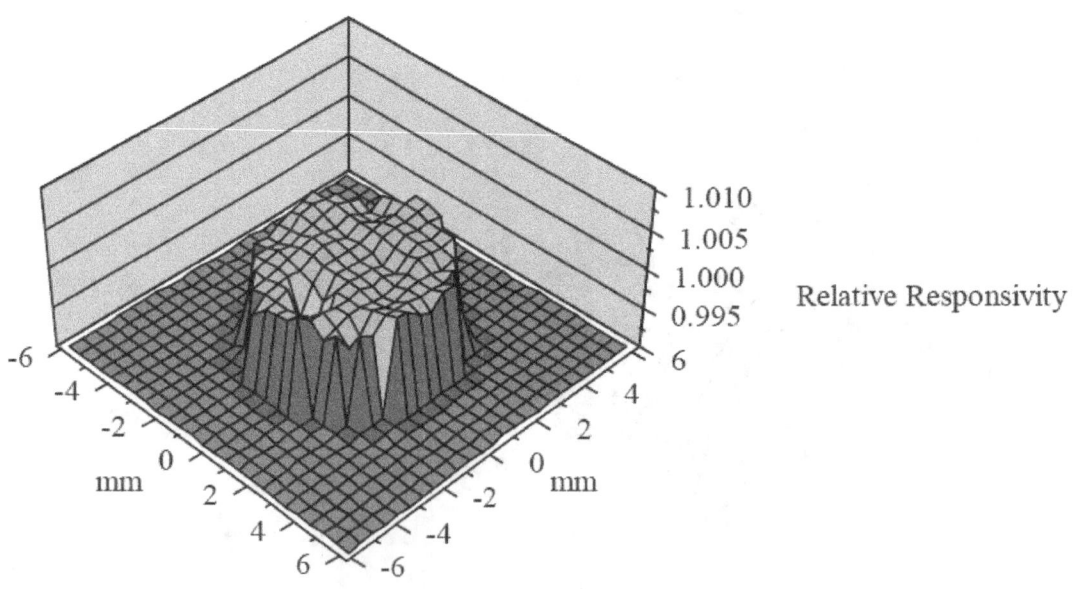

Test Date: December 24, 1997
NIST Test No.: 844/yyyyyy-97

A-24

REPORT OF TEST
NIST Test # 39081S - Responsivity Spatial Uniformity
Bigtime Government National Laboratory

Manufacturer: Acme Instruments
Model #: aa
Serial #: bbb

Figure 3a
Responsivity Uniformity of Acme Instruments aa, S/N bbb
0.2 % contours at 600 nm; 1.1 mm resolution; 0.5 mm/Step

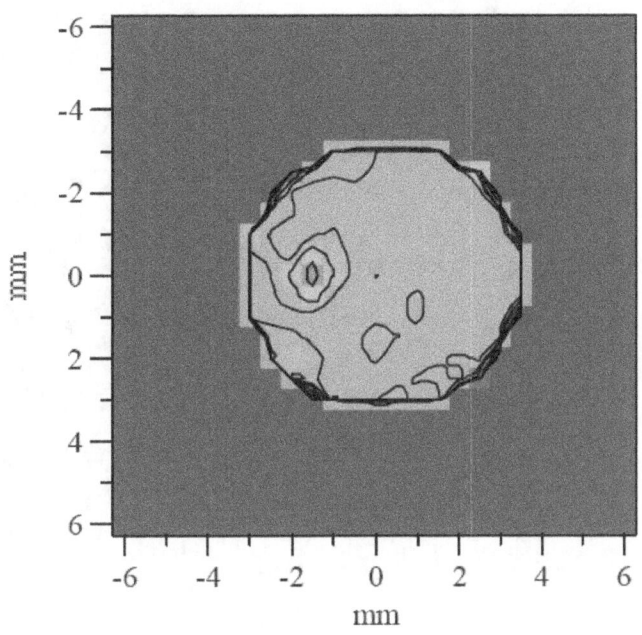

Figure 3b
Surface Plot of Responsivity Relative to
Center of Photosensitive Area for Acme Instruments aa, S/N bbb
at 600 nm; 0.5 mm/Step

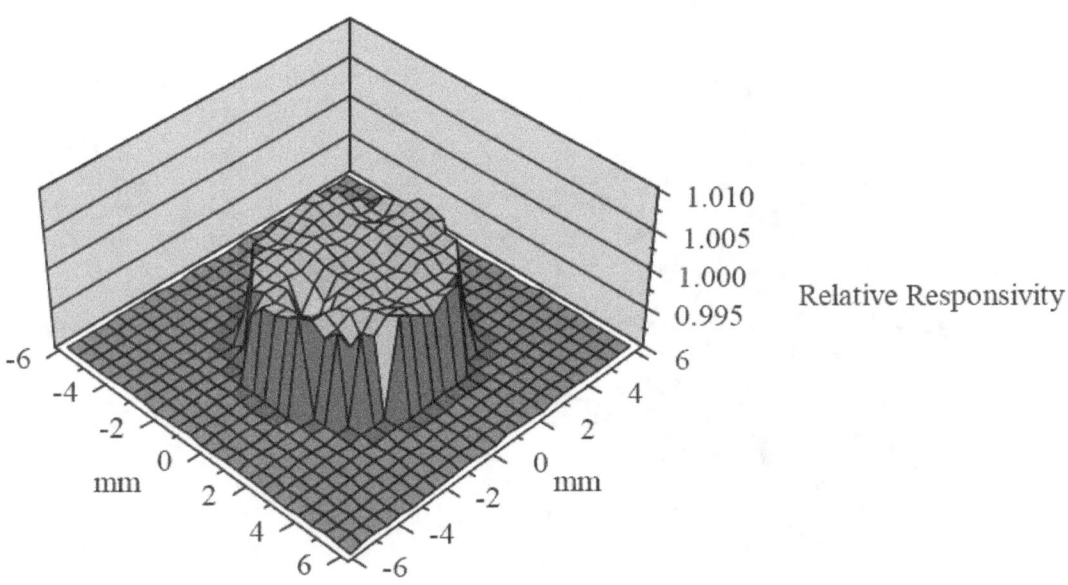

Test Date: December 24, 1997
NIST Test No.: 844/yyyyyy-97

REPORT OF TEST
NIST Test # 39081S - Responsivity Spatial Uniformity
Bigtime Government National Laboratory

Manufacturer: Acme Instruments
Model #: aa
Serial #: bbb

Figure 4a
Responsivity Uniformity of Acme Instruments aa, S/N bbb
0.2 % contours at 700 nm; 1.1 mm resolution; 0.5 mm/Step

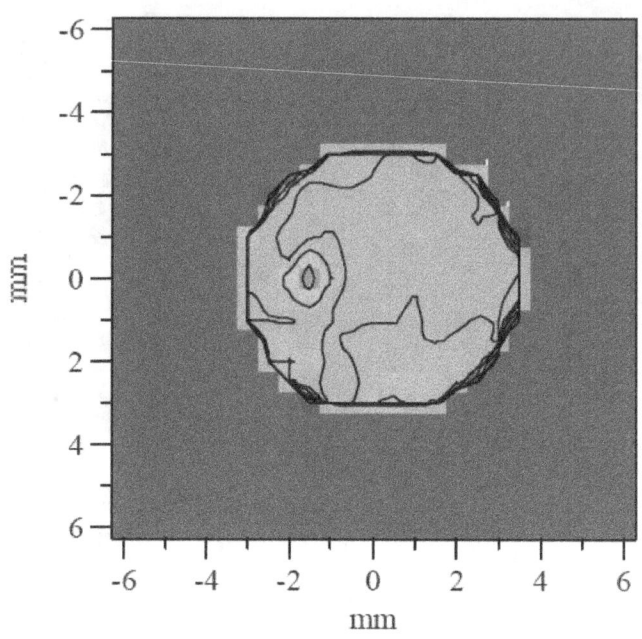

Figure 4b
Surface Plot of Responsivity Relative to
Center of Photosensitive Area for Acme Instruments aa, S/N bbb
at 700 nm; 0.5 mm/Step

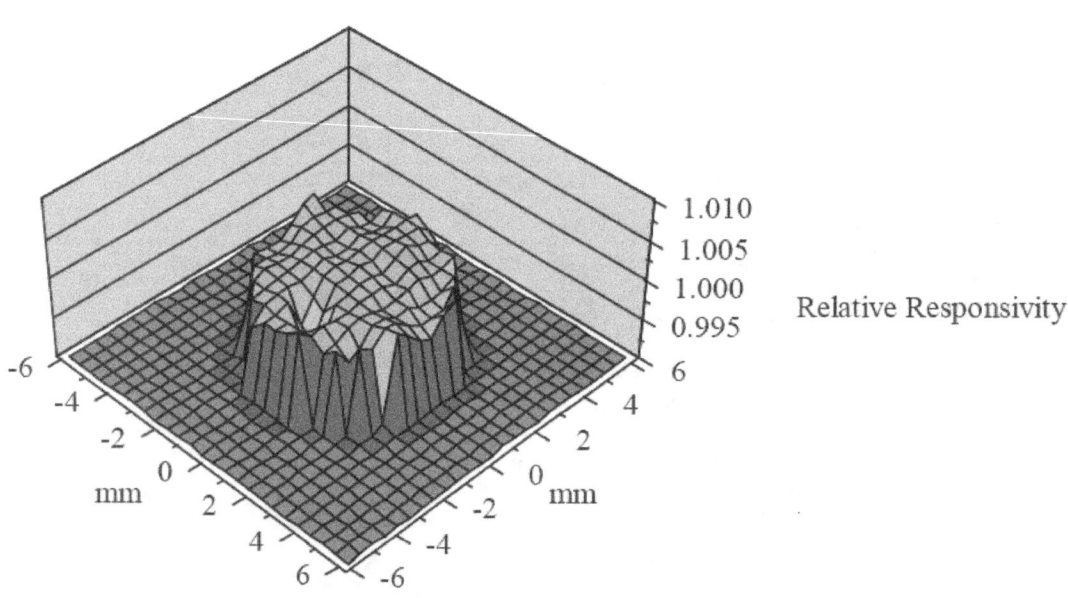

Test Date: December 24, 1997
NIST Test No.: 844/yyyyyy-97

www.ingramcontent.com/pod-product-compliance
Lightning Source LLC
Chambersburg PA
CBHW080302180526
45167CB00006B/2632